Reading in Science Workbook

Grade 5

Table of Contents

The activity sheets bearing the Wildlife Conservation Society logo are adapted from the Habitat Ecology Learning Program, a Wildlife Conservation Society curriculum designed by the Education Department at the Bronx Zoo. For more information on the society's curricula, teacher workshops nationwide and free National Teacher Membership program, visit www.wcs.org.

Cover Photo: Chris Johns/National Geographic; fern border—ThinkStock/Superstock.
Title page: ThinkStock/Superstock

The McGraw-Hill Companies

Printed in the United States of America
12 079 10 09

Life Science

Characteristics of Living Things

CHAPTER 1 • Classifying Living Things

CHAPTER 2 • Plant Structure and Functions

CHAPTER 3 • Plant Diversity

CHAPTER 4 • Animal Diversity

Life Science

UNIT B

Living Things and Their Environments

CHAPTER 5 • Interactions of Living Things

CHAPTER 6 • Ecosystems

Earth and Its Resources

CHAPTER 7 • Landforms, Rocks and Minerals

CHAPTER 8 • Air, Water and Energy

Astronomy, Weather and Climate

CHAPTER 9 • Astronomy

CHAPTER 10 • Weather

CHAPTER 11 • Weather Patterns and Climate

Physical Science

Properties of Matter and Energy

CHAPTER 12 • Properties and Structure of Matter

CHAPTER 13 • Forms of Matter and Energy

Physical Science

Motion and Energy

CHAPTER 14 • Newton's Laws of Motion

CHAPTER 15 • Sound Energy

CHAPTER 16 • Light Energy

Name ______________________ Date __________

Ungulate Criss-Cross

Read the paragraph about herbivores with hooves. The underlined words have already been used in the Criss-Cross. Read the directions below to find out how to fill in the rest of the blanks.

Ungulates are mammals with hooves. Most can be found grazing on different grasses. Some ungulates are ruminant animals that chew their cud. On the savanna in Africa, many ungulates have long legs and can run fast. This is very important because they are often hunted by the other wild animals that live with them on the plains.

*In the boxes below you'll find more words relating to African ungulates. Try to fit each of these words into its correct place in the crossword puzzle. Each word has only **one** correct place. Hint: First count the number of spaces available for a word in the puzzle. Then see what letters have already been entered and find the one word from the correct box that fits exactly. Start with those words that are already partly filled in and continue from there. It is a good idea to cross off the words in the boxes as you use them, so that you will not use them again by mistake.*

4 letters	5 letters	7 letters	8 letters	10 letters	11 letters	16 letters
mane	large	giraffe	antelope	rhinoceros	camouflaged	Thomson's
tail	nyala	stripes	blesboks	wildebeest		gazelles
horn	small		gemsboks			
	spots					
	zebra					

		W	I	L	D		H				P								
	R						U	N	G	U	L	A	T	E	S				
C	U	D					N				A				A				
	M						T				I				V				
	I						E				N				A				
	N						D				S				N				
	A							R							N		A		
	N							U							A		F		
	T					L	O	N	G	L	E	G	S				R		
								F					G	R	A	Z	I	N	G
M					G			A									C		
A					R			S									A		
M					A			T											
M					S												H		
A					S												O		
L					E												O		
S					S												V		
																	E		
																	S		

Name ______________________________ Date __________

Desertification Mix-Up

The paragraphs below describe important causes of desertification around the world. Fill in the correct word to fit in each blank from the bottom of each box.

Overgrazing

Cows, sheep, goats, and other __________ eat many plants that hold the soil in place. Then when it rains or floods, the precious soil can wash away. Wind can also blow away the soil. This is called "soil __________" and it is a major cause of desertification. Without __________, most plants cannot grow. Grazing livestock also __________ the ground and make it very hard. Then when it rains, the water runs off the soil instead of __________ in. The goat is called the "father of the desert" in Africa because it has caused so much damage.

TOPSOIL EROSION LIVESTOCK SOAKING TRAMPLE

Deforestation

Forests are cut down for farms, towns, and areas to __________ livestock. Trees are also cut down for __________ and paper products. When the trees are gone, the soil is exposed to harsh conditions. Wind and __________ can carry the soil away if there are no __________ to hold it in place. Nutrients and water are lost, and the area becomes more like a non-living __________.

TREES LUMBER DESERT GRAZE WATER

Collecting Wood

Millions of people have no oil or gas, so they rely on __________ to burn for fuel. In some dry areas too many plants are taken for __________. When the plants are removed, the soil is left __________. Then soil erosion becomes a problem. Sometimes people also break pieces off __________ trees, which can kill them. Then the plants and animals that depend on the trees for food or __________ disappear.

FUEL WOOD SHELTER LIVING BARE

Farming

When farmers grow crops, the soil loses __________ (which are like soil vitamins). To keep the soil rich, the land needs to rest. Today there are so many people trying to grow food that the land often doesn't have a __________ enough rest period. Another problem with farming is that people sometimes plant crops on hills or leave the soil bare for part of the year. Then the __________ washes away the soil. Irrigation __________ is another problem if the water does not drain off the field. When the water dries up, it leaves the salts that were in the water behind. Over time, the soil becomes too __________ to grow crops. Poisons used to kill weeds and insects can also harm the soil.

RAIN SALTY NUTRIENTS LONG WATERING

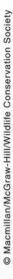

Name ______________________ Date __________

WILDLIFE CONSERVATION SOCIETY

Desert Specialist

Answer the questions below. Use the camel pictures to guide you.

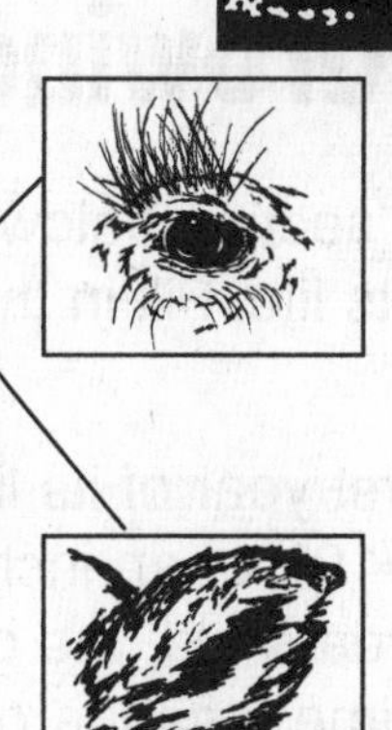

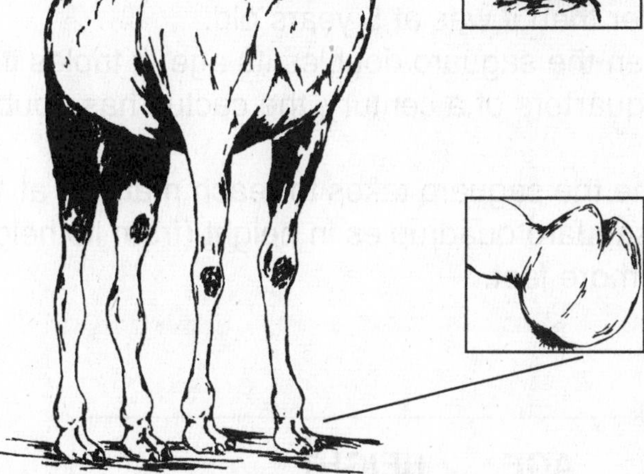

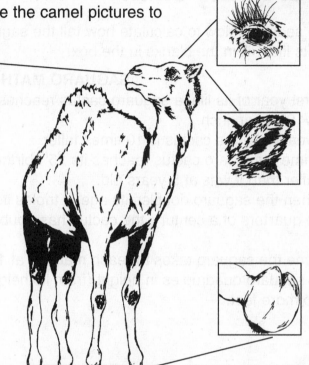

Dromedary Camel

1. Most people recognize the camel by its hump. What is in the camel's hump? ______________________
2. The camel's upper lip has a groove extending from each nostril to the mouth. What do you think is the purpose of this groove? (Hint: Think about water conservation.) ______________________
3. Since a camel can go up to eight days without water, it must be able to drink a lot at once. How many gallons of water can a camel drink at one time? ______________________
4. If you were walking with a camel through the desert, who do you think would sweat first? ______________________
5. Why does a camel need a hairy coat in the desert? ______________________
6. Why are the camel's wide feet a good adaptation for living in the desert? ______________________
7. What features on the camel's face would protect it during a sandstorm? ______________________

Name ______________________________ Date __________

WILDLIFE CONSERVATION SOCIETY

Slow Growing Giant

Read the section below to calculate how tall the saguaro is at each stage of its life. Fill in the blanks in the box.

SAGUARO MATH

- In the first year of its life, a saguaro cactus reaches a height of about 1/10th of an inch.
- At five years old, the cactus is 10 times taller.
- By the time a saguaro cactus reaches its 25th birthday, it's 24 times taller than it was at 5 years old.
- Then when the saguaro doubles its age, it triples its height.
- At three quarters of a century, the cactus has doubled its height again.
- In the time the saguaro takes to reach maturity at 150 years old, the saguaro quadruples in height (from its height at age 75) plus two more feet.

AGE	HEIGHT
1st year =	1/10 of an inch
5th year =	__________ inch(es)
25th year =	__________ feet
__________th year =	__________ feet
__________th year =	__________ feet
150 years =	__________ feet

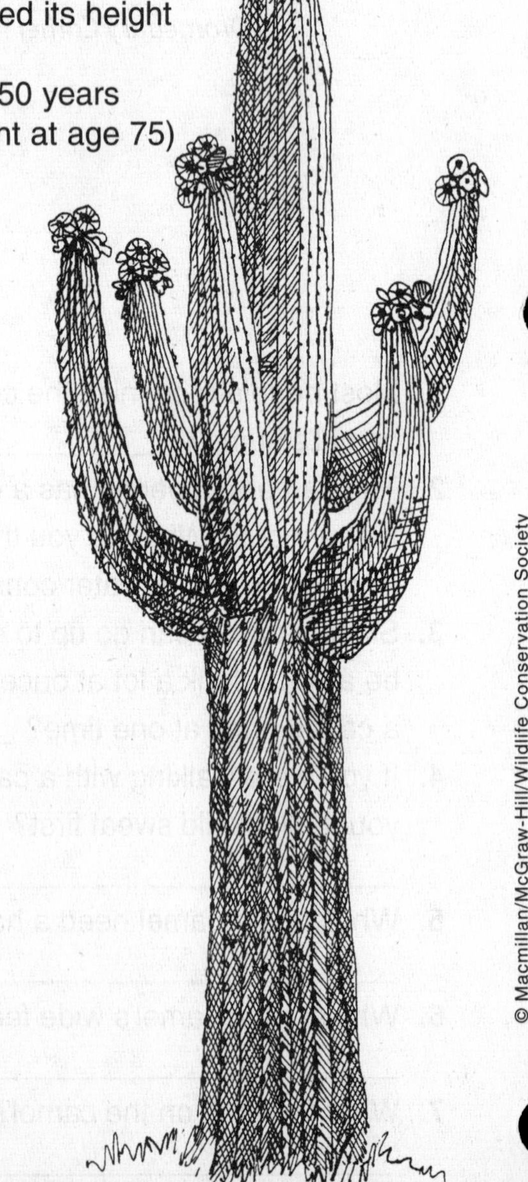

Name ______________________ Date ________

WILDLIFE CONSERVATION SOCIETY

Grasslands Climate: Mwanza, Tanzania

Study the graphs below. Then answer the questions on the bottom of the page.

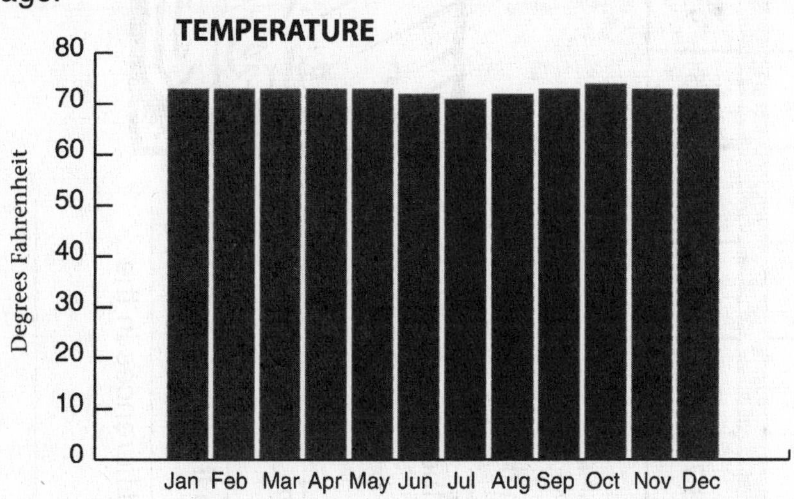

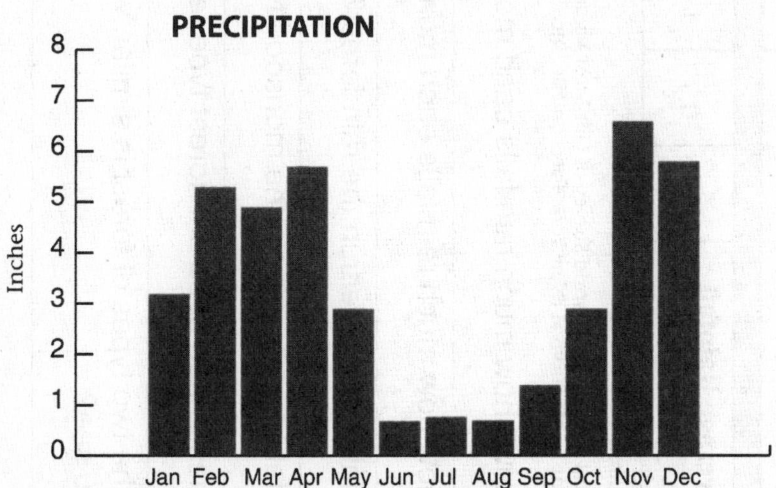

1. Which month has the most precipitation? ______________
2. Which three months have the least precipitation? ______________
3. Does the temperature change much during the year? ______________
4. What is the approximate temperature during the year? ______________
5. Does the grassland have seasons? If no, why not? If yes, what are they? ______________

Name ______________________ Date __________

WILDLIFE CONSERVATION SOCIETY

Wetter Where and When?

The data below show the pattern of rainfall in a tropical rain forest and a monsoon forest.

		Jan.	Feb.	Mar.	Apr.	May	June	July	Aug.	Sept.	Oct.	Nov.	Dec.
Rain Forest* (Uaupes, Brazil)	**Average Rainfall** (inches)	10.3	10.8	12.9	10.0	13.0	9.6	10.0	7.3	6.5	7.9	9.1	13.1
Monsoon Forest** (Cochin, India)	**Average Rainfall** (inches)	0.7	0.7	1.7	3.7	11.5	27.5	25.7	12.5	9.5	12.5	7.0	1.7

* Data courtesy of NOAA National Climatic Data Center

** Data from: Arnold Newman, Tropical Rainforest, A World Survey of Our Most Valuable and Endangered Habitat with a Blueprint for Its Survival (New York: Facts on File, Inc., 1990), p. 24.

1. Draw a graph that shows how much rain falls each month in the rain forest?

2. Draw a graph that show how much rain falls each month in the monsoon forest?

3. How much total rain falls in one year in the rain forest?

4. How much total rain falls in one year in the monsoon forest?

5. How many more inches of rain does one forest type receive than the other?

6. If the total rainfall in the two types of forest is similar, what differences in the rainfall pattern do you see?

7. How could this difference affect the plants or animals that live in the two forests?

Name ______________________ Date __________

WILDLIFE CONSERVATION SOCIETY

Rain Forest Poems

Follow the steps below and write a poem about a rain forest animal. Use your imagination!

1. ________

2. ________ ________

3. ________ ________ ________

4. ________ ________ ________ ________

5. ________

How to write poetry using the Cinquain style

1. Use one word (noun) to name the object you are writing about.
2. Use two words (adjectives) to describe #1.
3. Use three words (verbs) to describe what #1 might be doing.
4. Use a four-word phrase to tell how you feel about #1.
5. Use another word (synonym) that means the same as #1.

1. ________

2. ________ ________

3. ________ ________ ________

4. ________ ________ ________ ________

5. ________

Macaw

Name ______________________________ Date __________

Where Are the Temperate Forests?

Use the map to create a chart. In one column, list the provinces and states that mainly have coniferous forests. In another column, list the states with deciduous forests.

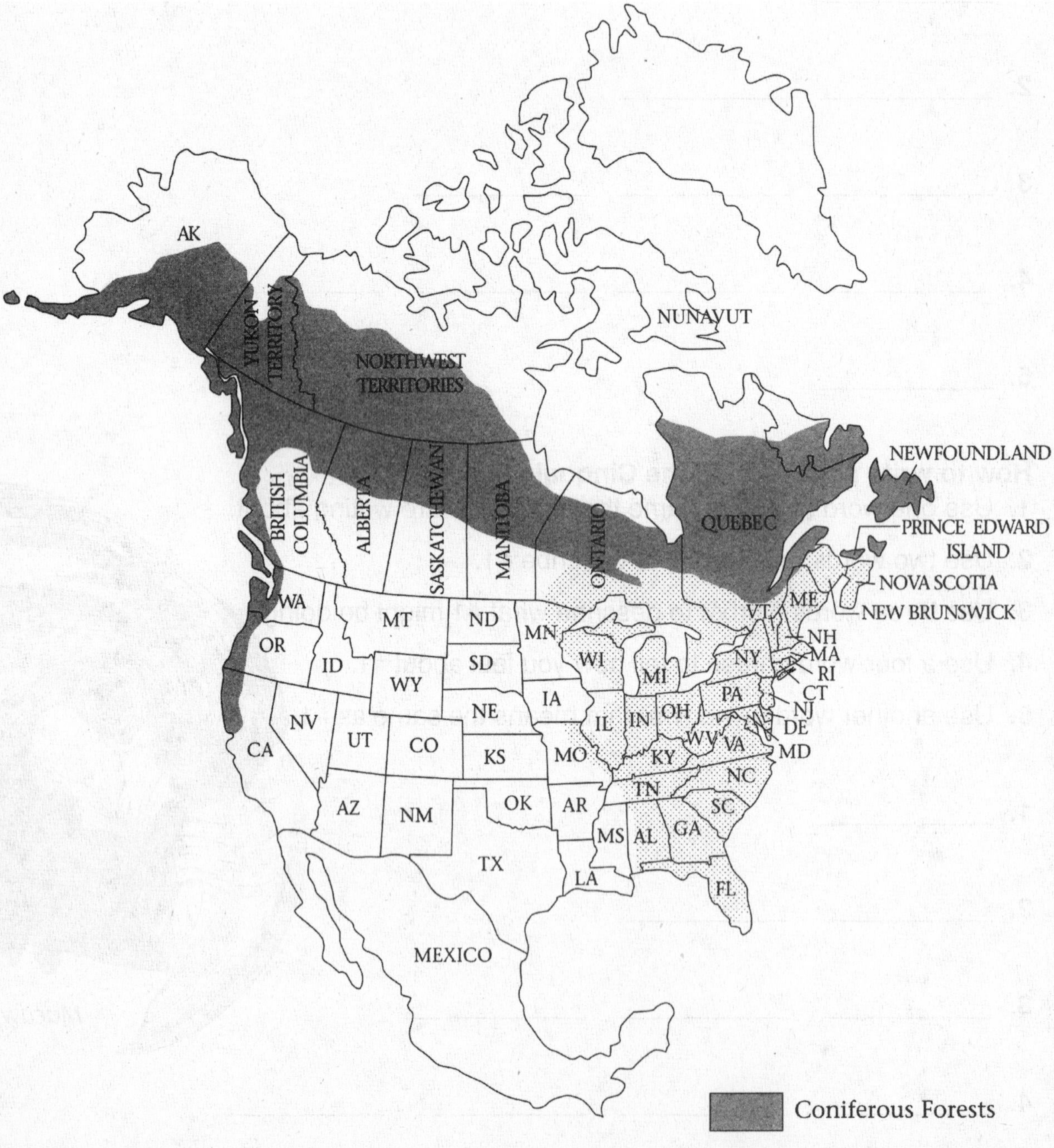

AK
YUKON TERRITORY
NORTHWEST TERRITORIES
NUNAVUT
BRITISH COLUMBIA
ALBERTA
SASKATCHEWAN
MANITOBA
ONTARIO
QUEBEC
NEWFOUNDLAND
PRINCE EDWARD ISLAND
NOVA SCOTIA
NEW BRUNSWICK
WA
OR
CA
NV
ID
MT
WY
UT
CO
AZ
NM
ND
SD
NE
KS
OK
TX
MN
IA
MO
AR
LA
WI
IL
MI
IN
OH
KY
TN
MS
AL
GA
FL
SC
NC
VA
WV
PA
NY
VT
ME
NH
MA
RI
CT
NJ
DE
MD
MEXICO

Coniferous Forests

Deciduous Forests

MACMILLAN
McGRAW-HILL

Science

Reading in Science

Name________________________________ Date______________

Chapter Graphic Organizer
Chapter 1

Classifying Living Things

The graphic below is a Venn diagram. A Venn diagram can be used to compare and contrast two items that have both similarities and differences. Use the diagram below to show how plant cells and animal cells are alike and how they are different. Use the words listed in the word bank to complete the diagram.

List the characteristics that plant and animal cells share in the intersecting part of the diagram. List the characteristics that apply to only animal cells in the circle labeled Animal Cells. List the characteristics that apply to only plant cells in the circle labeled Plant Cells.

Word Bank

cell membrane cytoplasm nucleus mitochondria cell wall chloroplasts

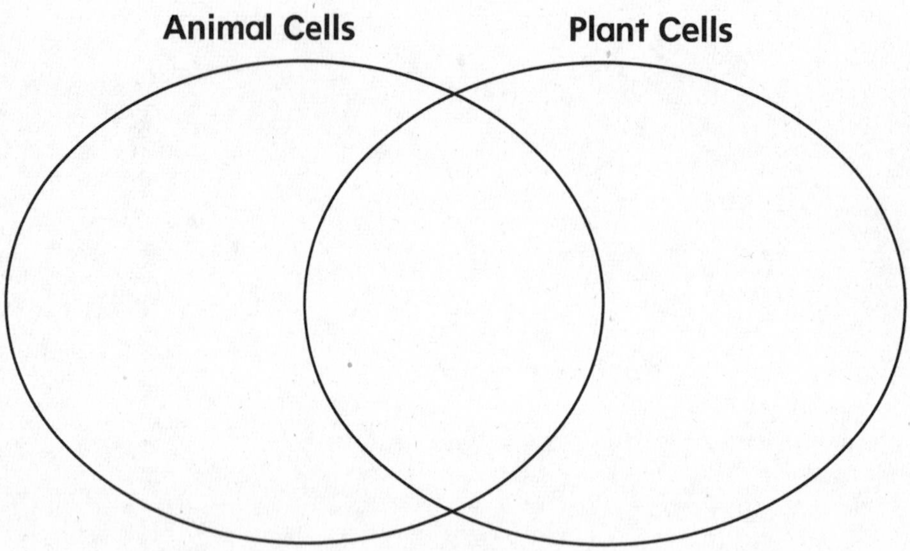

Animal Cells

Plant Cells

Draw Conclusions

Read each set of facts, or clues, in the boxes below. Then draw and label your conclusions in the spaces provided.

I am found in plants. I contain chlorophyll that plants use to make their own food. What am I?	I am the basic unit of life. I am found in plants and in animals. What am I?
I can be one cell or I can have many cells. I absorb food from decaying dead organisms and wastes in my environment. What am I?	I am a single-celled organism. I do not have a nucleus. I cause diseases in plants and animals. What am I?

Draw Conclusions

When you draw conclusions, you decide if something is true. To do that, you need all the facts. Sometimes you have them, but sometimes you do not. If you don't have all the facts, you can't draw a conclusion.

Read the first four sets of facts below. Can you draw a conclusion? Circle your answers. Read the last two sets of facts and write your answers.

1. Plant cells have cell walls to support the plants. Trees have cell walls. Are trees plants?

   yes　　no　　not enough facts

2. Some fungi can spoil food and make people sick. Kevin was sick last week. Did a fungus make him sick?

   yes　　no　　not enough facts

3. Vascular plants have tube like cells that carry water from the roots to the rest of the plant. Mosses do not have this kind of tissue. Are mosses vascular plants?

   yes　　no　　not enough facts

4. The protist kingdom includes living things that can be seen with a microscope. It also includes living things that can be seen without a microscope. Paula saw some seaweed. Is the seaweed a protist?

   yes　　no　　not enough facts

5. Carly has twin cousins, Maria and George. She also has a sister, Laura. Carly is three years older than Maria and George and one year younger than Laura.

   a. If Carly is 11, how old is Laura?

   _______________________________________.

   b. How old are Maria and George? _______________.

6. Joey gets home from school an hour earlier than his older sister, Lisa. Lisa gets home from school three hours later than their younger sister Lucy.

   a. If Joey gets home at 2:30, what time does Lisa get home?

   _______________.

   b. What time does Lucy get home? _______________.

Name ______________________ Date ____________

Lesson Outline
Lesson 1

The Basic Unit of Life

Fill in the blanks.  Reading Skill: **Draw Conclusions** - 3, 15

What Is the Basic Unit of Life?

1. An organism's basic structure of life is called a(n) ________________.
2. Plant cells contain a green substance called ________________.
3. Because of chlorophyll, green plants can use the ________________ to make food.
4. Plant cells have rigid ________________ to support them.
5. These parts of a cell, present in plant cells, but not in animal cells, are called the ________________ and ________________.
6. The ________________ helps supply energy from the cell.

What Are Living Things Made of?

7. The parts of an organism that have a major role in helping living things function properly are called ________________.
8. Water and food are ________________ into plant cells through roots, trunks, stems, and branches.
9. Long threadlike cells that help animals move their bodies are called ________________.
10. Group of cells are organized into ________________.
11. The ________________ help carry water and food from the roots to the leaves of celery.

Name_________________________________ Date_______________

Lesson Outline

Lesson 1

What Traits Are Used to Classify Organisms?

12. People use ____________________ to help us understand our surroundings.
13. The science of finding patterns of living things is called ____________________.
14. Scientists classify organisms by studying the ____________________.
15. Early systems for classifying plants were based on characteristics people could see easily, because ____.
16. Aristotle classified plants into three groups based on ________________.

Name ______________________ Date ____________

Interpret Illustrations

Lesson 1

Plant and Animal Cells

Diagrams like these show the different parts of something—in this case, plant cells and animal cells. In studying this type of diagram, the first step is to locate each feature or structure and read its label. When you have located a feature, remember what you have read about what it does. Notice its relationship to other features. If you have two diagrams side by side, look for similarities and differences between them.

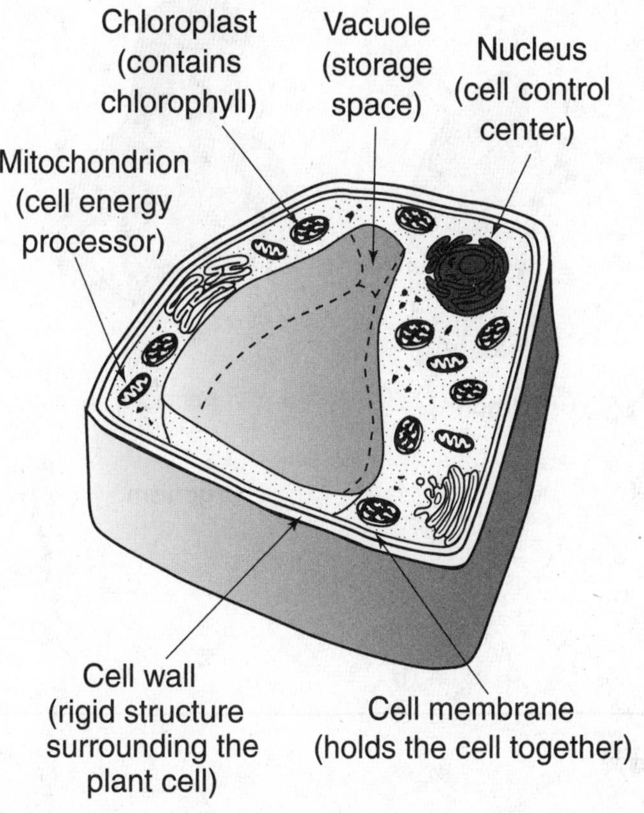

Plant cell

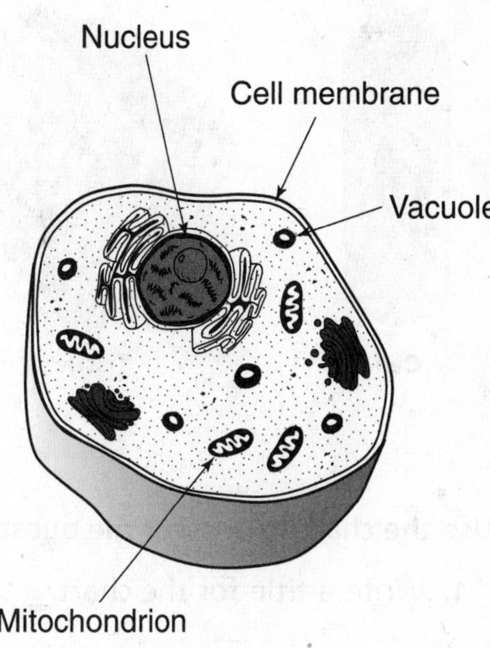

Animal cell

Answer these questions about the diagram above.

1. What is the control center in plant cells called? ____________________
2. The large storage area inside a plant cell is called a(n) ____________________
3. What is the rigid structure surrounding a plant cell called?

   ____
4. Does an animal cell have this structure? ____________________
5. What other feature does a plant cell have that an animal cell lacks?

   ____

Name ______________________ Date ____________

Interpret Illustrations

Lesson 1

What Are Living Things Made of?

The chart shows the level of organization of many-celled living things. The level of organization increases when the chart is read from left to right. Each level of organization includes all the levels to the left in the chart.

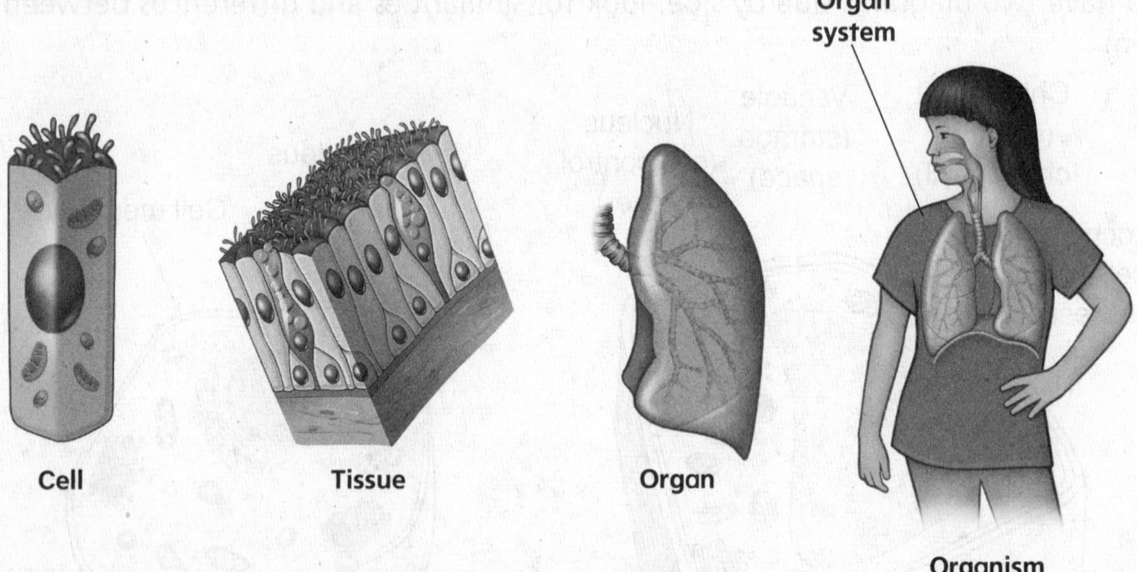

Use the chart to answer the questions.

1. Write a title for the chart.

____

2. What are organ systems made up of?

____

3. Which has a higher level of organization: a cell or an organ?

____

4. From the chart, which is the most highly organized?

____

5. How is a tissue like an organ? How are they different?

____

____

____

Name ______________________ Date ____________

Lesson Vocabulary
Lesson 1

The Basic Unit of Life

Vocabulary

- chlorophyll
- chloroplast
- tissue
- organ
- organ system

Fill in the blanks.

1. Similar cells that have the same job or function come together to make a(n) ______________.
2. A(n) ______________ is a group of organs that work together to do a certain job.
3. The green chemical found in many plants is ______________.
4. Tissues of different kinds come together to make an ______________.
5. A leaf makes food in cells that contain ______________.

Answer the following questions.

6. What is the difference between animal and plant cells?

____

____

7. How do cells come together to make organ systems?

____

____

8. Name two plants whose leaves we use for food, two plants whose stems we use for food, and two plants whose roots we use for food.

____

____

____

Name ______________________ Date ____________

Cloze Test
Lesson 1

The Basic Unit of Life

Vocabulary

animal chloroplasts cells organisms different classification

Fill in the blanks.

All living things are made of ____________. Humans are complex ____________ that have different cells that have different functions. The cells of ____________ organisms are different. Plant cells are different from ____________ cells. Plant cells have ____________ and cell walls. Animal cells do not. The science of finding patterns among living things is called ____________.

The Kingdoms of Life

Fill in the blanks.  Reading Skill: **Draw Conclusions** - 4, 7, 11, 12

What Traits Are Used to Classify Plants?

1. The largest subdivision of living things is called a ______________.
2. Plants with tubes that help the plant carry water and food are called ______________.
3. Vascular plants can grow taller and thicker than mosses because they have ______________.
4. Plants without tubelike structures to carry water are called ______________.
5. Two types of nonvascular plants are ______________ and ______________.
6. Vascular and nonvascular plants are known as the two types of ______________ in the plant kingdom.

What Makes Animals Different from Plants?

7. Two major factors that make animals different from plants are:
   a. ______________________________ and
   b. ______________________________.
8. The animal kingdom is divided into two large groups called ______________ and ______________.
9. Phyla can be broken down into smaller divisions called ______________.

What Is a Fungus?

10. Fungi don't make their own food or eat animals for food, but instead ____________________ food from decaying dead organisms and wastes in its environment.

What Is a Protist?

11. The common characteristic of members in the protist kingdom is ____.

12. Protists can be found anywhere there is ____________________.

What Are Bacteria?

13. A microscopic organism consisting of one cell without a nucleus is a(n) ______________________________.

14. The kingdom of bacteria was once known as ____________________.

Name ____________________ Date __________

Interpret Illustrations
Lesson 2

What Are the Major Plant Groups?

Charts like the one below are a way of organizing, or classifying information. They are like a visual outline—they start with the most general topic or group and break it down into smaller and smaller divisions. One clue to reading this kind of chart is the use of bold (dark) print. Bold print is used for the largest grouping in the chart: Plant Kingdom. Lighter letters are used to label smaller divisions.

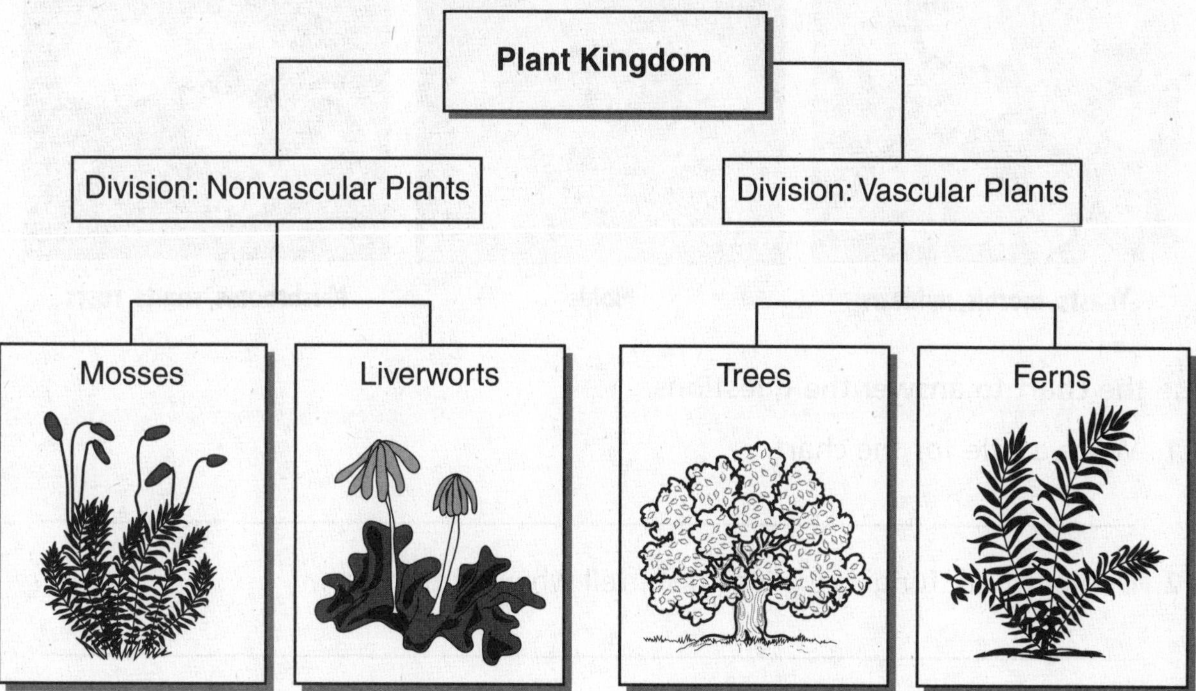

Answer these questions about the diagram above.

1. The plant kingdom is divided into two major groups. They are:

   ____

2. In which major group do trees belong? ____________________
3. In which major group do ferns belong? ____________________
4. From the chart, give one example of a nonvascular plant.

   ____

5. What is the only grouping in this chart that includes both trees and mosses?

   ____

Name______________________________ Date____________

Interpret Illustrations

Lesson 2

What Is a Fungus?

The chart below shows the groups of the Fungus Kingdom. Look at each picture to see the differences between each type of mushroom.

Fungus Kingdom

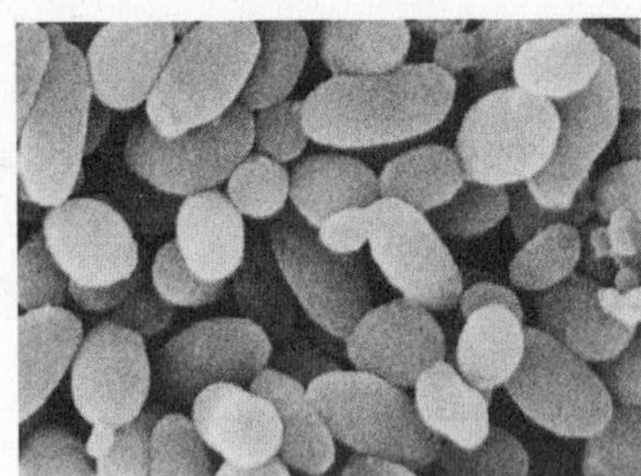

Yeasts, morels, mildews

Molds

Mushrooms, smuts, rusts

Use the chart to answer the questions.

1. Write a title for the chart.

   ____

2. Do you think fungi are alive? If so, tell why you think so.

   ____

3. Which type of fungus has its own group? ____________________

4. Name one way humans use helpful fungi.

   ____

5. Describe one way fungi can make people sick.

   ____

Name____________________ Date__________

Lesson Vocabulary

Lesson 2

The Kingdoms of Life

Vocabulary

vascular
nonvascular
invertebrate
vertebrate
fungus
protist
bacteria

Fill in the blanks.

1. An organism that absorbs food from dead organisms and waste is called a(n) ________________.
2. A moss is an example of a(n) ________________ plant.
3. Animals that have backbones are called ________________.
4. Algae and other tiny organisms that live in water are classified as ________________.
5. Animals that do not have backbones are called ________________.
6. Plants that have tubelike structures up and down the stems are ________________ plants.
7. The tiniest living things are ________________.

Answer each question.

8. What is the major difference between vertebrates and invertebrates?

____

____

9. What is the difference between vascular plants and nonvascular plants?

____

____

10. List two functions that fungi have.

____

____

____

Name____________________ Date__________

Cloze Test
Lesson 2

The Kingdoms of Life

Vocabulary

food	waste	many	flu-like
rise	cheeses	fungus	diseases

Fill in the blanks.

A fungus can't move around or make its own ________________. Fungi absorb food from decaying organisms and ________________ in the environment. A fungus can be one celled or ________________ celled. Some fungi contain chemicals that fight ________________. Others keep our environment clean. A fungus turns ________________ sharp and tangy. Yeast, another fungus, helps bread to ________________. Certain fungi produce problems. Athlete's foot, an itchy disease, is another ________________. Others that grow in damp places like basements cause ________________ symptoms.

Name______________________________ Date____________

Chapter Vocabulary

Chapter 1

Classifying Living Things

Circle the letter of the best answer.

1. A plant that has tubelike structures up and down the stem is a(n)
   a. fungus. b. protist.
   c. bacteria. d. vascular plant.

2. The green food factories of plants are called
   a. stomata. b. petioles.
   c. chloroplasts. d. veins.

3. The substance that uses the Sun's energy to make food for plants is
   a. chlorophyll. b. minerals.
   c. spores. d. water.

4. Plants that have tissue through which food and water move are said to be
   a. bacteria. b. nonvascular.
   c. protist. d. vascular.

5. A group of organs that work together to do a certain job is a(n)
   a. organ system. b. tissue.
   c. brain. d. cell.

6. Animals that do not have backbones are
   a. vertebrates. b. mammals.
   c. jellyfish. d. invertebrates.

7. A tissue is a group of
   a. similar cells that have the same job function.
   b. different organs that come together.
   c. invertebrates.
   d. vertebrates.

Name________________________________ Date______________

Chapter Vocabulary

Chapter 1

Circle the letter of the best answer.

8. Animals that have backbones are

a. vertebrates. **b.** reptiles.

c. agnatha. **d.** invertebrates.

9. A group of tissues that come together is a(n)

a. organ. **b.** system.

c. cell. **d.** building block.

10. The green chemical that allows plants to make their own food is found in

a. vacuoles. **b.** chloroplasts.

c. cell membranes. **d.** mitochondria.

11. Mushrooms, molds, and rusts are examples of

a. ancient bacteria. **b.** protists.

c. true bacteria. **d.** fungi.

12. Diatoms, slime molds, and green algae are

a. ancient bacteria. **b.** protists.

c. euglenas. **d.** dinoflagellates.

13. Animal cells do not have

a. cell membranes. **b.** cell walls.

c. nuclei **d.** mitochondria.

Name ____________________ Date ____________

Chapter Summary

Chapter Summary

1. What are four vocabulary words you learned in the chapter? Write a definition for each.

2. Which diagram helped you to understand an idea better?

3. What are two main ideas that you learned in this chapter?

Name_________________________________ Date_______________

Chapter Graphic Organizer
Chapter 2

Plant Structure and Functions

In the Ideas Circle below, plant parts are shown in the boxes on the top. Things plants do to survive are shown in the boxes on the bottom. Draw an arrow from the name of a plant part to one way the part helps the plant survive.

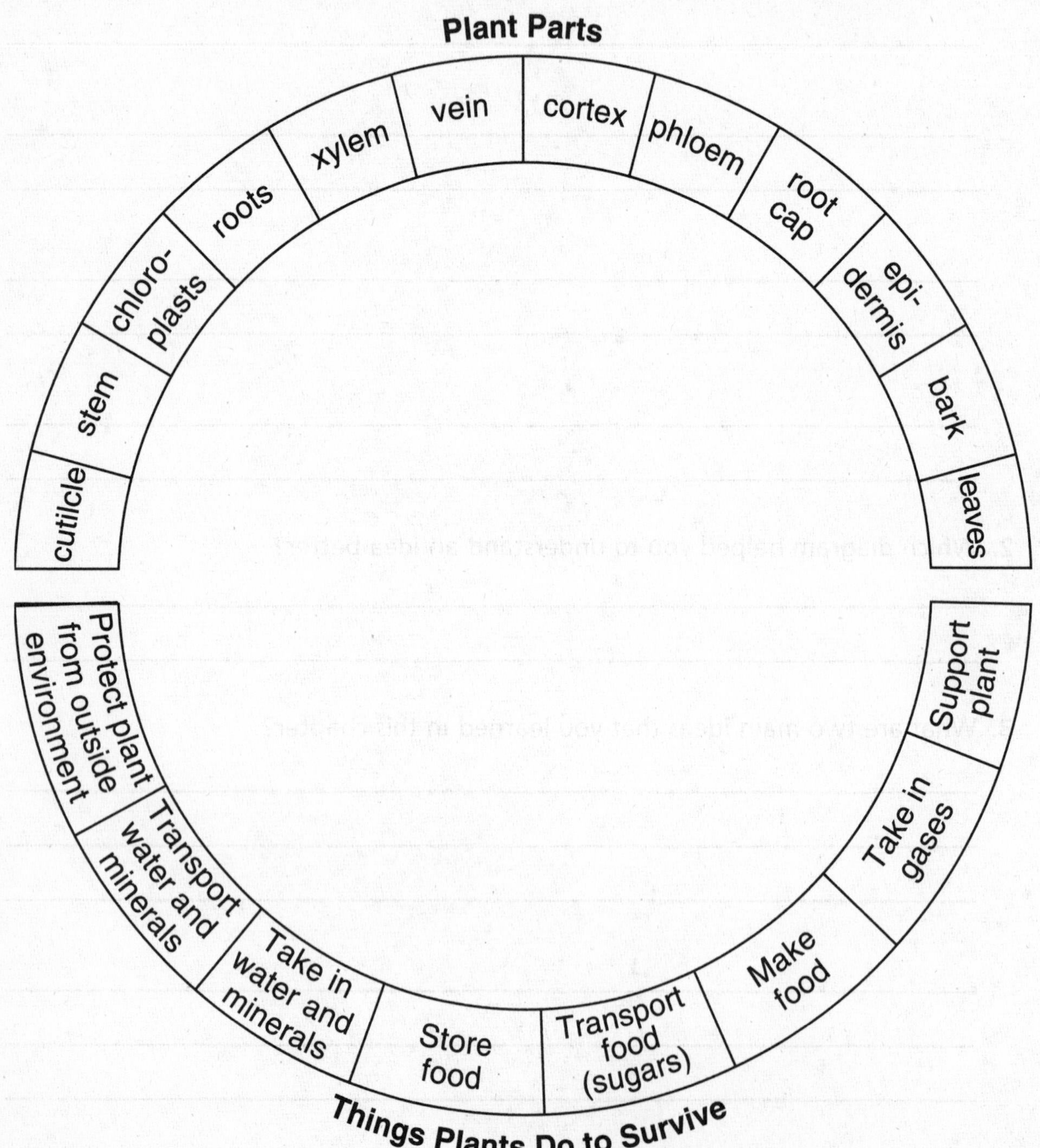

Name_________________________________ Date_______________

Chapter Reading Skill
Chapter 2

Main Idea and Supporting Details

Every article has a main idea. The supporting details give information about the main idea.

Read each article below based on information from this chapter. Then read the list of suggested main ideas, and circle the correct one.

1. All plants need water to grow. Water travels from the roots to the leaves of plants. Minerals and water from the soil enter through the plant's root hairs. The water passes through the cortex of the root, enters the xylem, and travels up the stem. Transpiration in the leaves helps draw the water into the xylem of the plant's stem. Water moves up the stem, through a leaf's petiole, and into its veins. The veins carry the water to the leaf's cells. Most of the water that entered the roots is given off into the air through the leaf's stomata during the process of transpiration. The water follows a path through all the parts of a plant.

The Main Idea:

a. Transpiration gives off most of the plant's water.

b. Water travels through all parts of a plant.

c. Minerals and water come from the soil.

2. Many of the foods we eat come from different plant parts. Cinnamon is a yellowish-brown spice made from the dried inner bark of certain trees. It adds a distinct flavor to pies and cakes. The maple syrup you pour over pancakes comes from the sap of the maple tree. Sweet potatoes, beets, carrots, horseradishes, and parsnips are all roots of plants. You can see the stems of celery but ginger, potatoes, and sugar are stems, too. Basil, cabbage, lettuce, onion, oregano, parsley, and spinach are leaves that can make a good salad. Peppermint leaves make a soothing tea. The flowers from broccoli and cauliflower are easy to spot. Chocolate, corn, peanuts, rice, and wheat are seeds that are used to make many different kinds of food. Apple, oranges, tomatoes are fruits that are probably a regular part of your diet. Can you think of other plant parts that you like to eat?

The Main Idea:

a. Different plant parts are easy to see.

b. Fruits and vegetables are food we eat every day.

c. People use different plant parts for food.

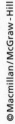

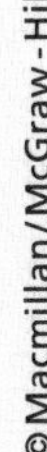

Name______________________ Date__________

Chapter Reading Skill

Chapter 2

Find the Subject

Read each paragraph based on information in your book. Then write the main idea for each paragraph.

1. The parts of a leaf work together to keep a plant alive. The epidermis is the outermost layer of a leaf. It secretes a waxy coat, called a cuticle, that keeps water from leaving the leaf. In cells between the layers of the epidermis the leaf makes food. Air comes through tiny pores in the bottom of leaves called stomata. Guard cells open and close the stomata. They make sure that the plant has plenty of water.

The Main Idea:

____

____

2. There are certain things that all stems have in common. All stems support leaves. Stems help leaves reach higher places to get sunlight. Without stems there would be no transportation system for water and minerals to move from the roots to all parts of the plant. Stems move food made in the leaves to all other parts of the plant.

The Main Idea:

____

____

3. Aside from supporting a plant and acting as a transportation system, some stems do more. Potatoes and sugarcane have stems that store food for the plants to use at a later time. Water is stored in the stems of cactus plants. This is extremely helpful when there are long dry periods in the desert. The stems of asparagus actually help make food.

The Main Idea:

____

____

4. All plants need sunlight, water, and nutrients to survive and grow. This means they must compete with other organisms in their environment. Each plant has its own way of competing. For example, some plants make chemicals that stop insects and other animals from feeding on or infecting them. Giant redwoods preserve the nutrients and water in the soil for themselves. They grow so tall that sunlight is blocked from reaching the ground. Without sunlight not many plants can grow near the giant redwoods.

The Main Idea:

____

____

____

Use with textbook pages A26–A53

Name ______________________ Date __________

Lesson Outline

Lesson 3

Roots, Stems, and Leaves

Fill in the blanks.  Reading Skill: **Draw Conclusions** - questions 6, 9, 18, 22, 25

How Do a Plant's Parts Help It Survive?

1. Roots help plants survive in three main ways:
   a. ______________________,
   b. ______________________, and
   c. ______________________.
2. A layer of tough cells that protect the growing tip of the root is called the ______________.
3. A large thick root with a few hairy branching roots is a(n) ______________.
4. Plants with fibrous roots thrive in dry regions because ______________________
5. Water and minerals pass through the root's ______________ and move upward in the ______________ to all the parts of the plant, while food flows downward in the ______________.

How Are Stems Similar?

6. Plant stems support the ______________ and ______________.
7. The transportation system in plant stems has three basic parts: ______________, ______________, and ______________.
8. Besides providing support and transportation, some plant stems are storage places for ______________ and ______________.

What Are Leaves?

9. The leaves of a maple tree, which hang singly, are ______________ leaves.
10. Leaves that grow in clusters are ______________ leaves.
11. The epidermis secretes a waxy coating called the ______________, which keeps water inside the leaf.

12. Besides sunlight, chloroplasts need three things to make food: ____________, ____________, and ____________.

13. Air comes through tiny pores in the bottom of the leaves called ____________.

14. Different leaf shapes, such as these, have different purposes:
    a. broad, flat green surfaces: ____________
    b. cactus spines: ____________
    c. Venus's-flytrap leaves: ____________

15. When water evaporates from the leaves, more water moves up through the plant to replace the lost water in a process called ____________.

What is Photosynthesis?

16. All living things need ____________ to survive.

17. The food-making process in plants is called ____________.

18. Plants and animals break up sugar, releasing energy, in the process of ____________.

How Does Water Get from Roots to Leaves?

19. Plants lose water through ____________ from their leaves.

20. Water and minerals from the soil enter through the plant's ____________.

What Parts of Plants Do You Eat?

21. Broccoli and cauliflower are the ____________ parts of plants.

22. When you eat peanuts and corn, you are eating ____________.

Name ______________________ Date __________

Interpret Illustrations

Lesson 3

What Is Photosynthesis?

The leaf in this diagram is like a small factory where the process of photosynthesis takes place. The arrows in the diagram show the substances that enter and leave the leaves of green plants during this process. Notice the formula on top of the diagram that sums up the process. The arrow in the formula shows the results of the chemical processes that take place in the leaf.

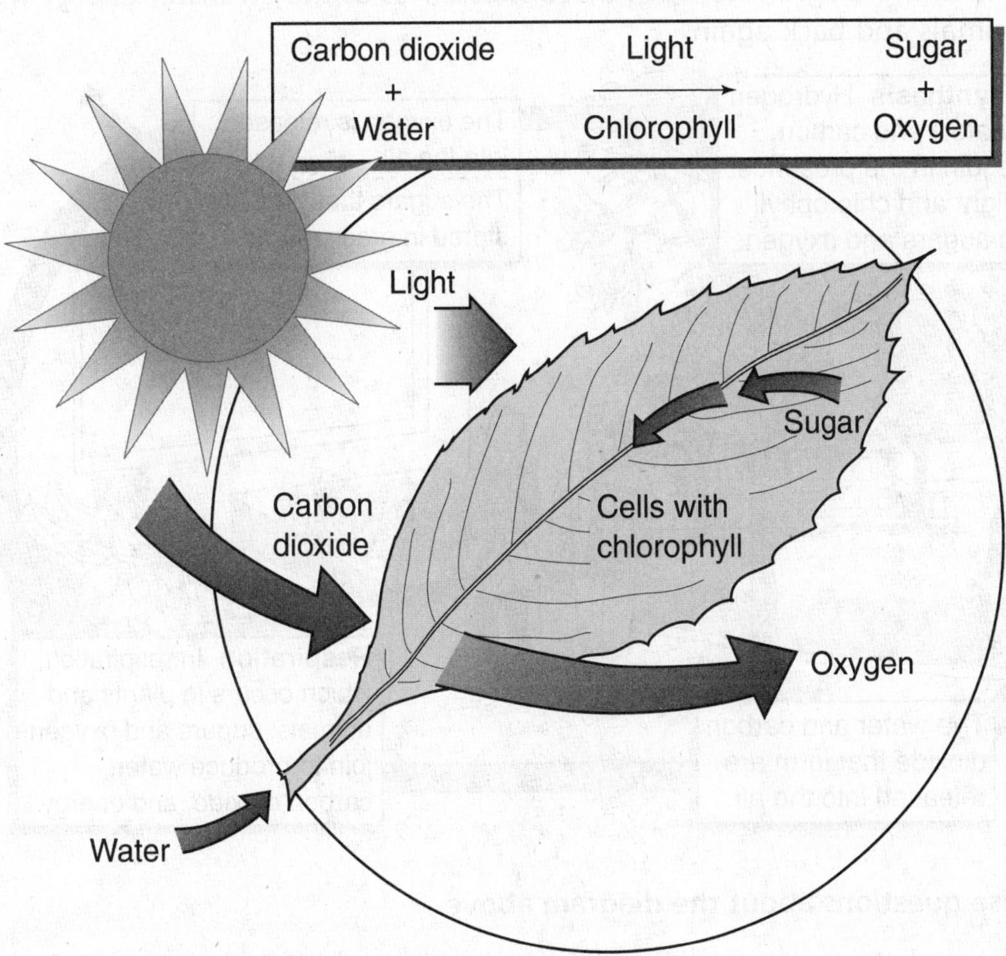

Answer these questions about the diagram above.

1. What is the source of the light used in photosynthesis? ______________.
2. What chemical enters the leaf from the air? ______________.
3. The chemical in the leaf cells is ______________.
4. The leaf makes food in the form of ______________.
5. What substance does the leaf give off into the air? ______________.

Name______________________________ Date__________

Interpret Illustrations

Lesson 3

Oxygen and Carbon Dioxide Exchange

When you see arrows in a diagram, you can expect that the diagram will show a process or cycle. The arrows indicate the direction of movement or change. This diagram shows the cycle by which plants and animals exchange oxygen and carbon dioxide. It involves two processes—photosynthesis and respiration. Follow the paths of the arrows in the diagram to track these substances as they transfer energy from plants to animals and back again.

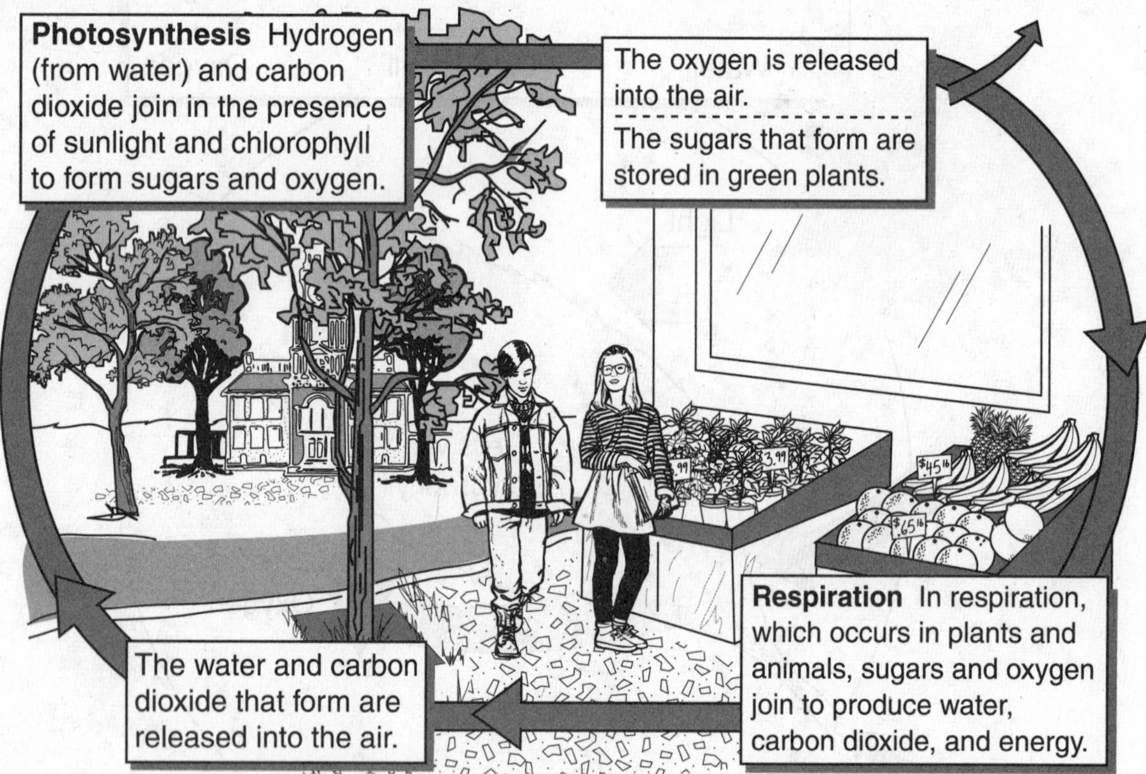

Photosynthesis Hydrogen (from water) and carbon dioxide join in the presence of sunlight and chlorophyll to form sugars and oxygen.
The oxygen is released into the air.
The sugars that form are stored in green plants.
Respiration In respiration, which occurs in plants and animals, sugars and oxygen join to produce water, carbon dioxide, and energy.
The water and carbon dioxide that form are released into the air.

Answer these questions about the diagram above.

1. What two substances are produced by green plants during photosynthesis?

   ________________________.

2. Animals breathe in the ________________ that is released into the air during photosynthesis.

3. During respiration, water, ________________, and ________________ are produced.

4. The process of ________________ occurs in both plants and animals.

Name ______________________ Date ____________

Lesson Vocabulary

Lesson 3

Roots, Stems, And Leaves

Vocabulary

- root cap
- phloem
- cambium
- cortex
- epidermis
- transpiration
- xylem
- root hairs

Fill in the blanks

1. Threadlike parts of cells on the surface of a root are called ____________.
2. A layer of roots and stems that stores food is the ____________
3. Water and minerals flow up from the roots through the ____________.
4. The ____________ is a layer in a root that separates the xylem from the phloem.
5. A ____________ protects the root tip as it grows into the soil.
6. The process of ____________ occurs when water vapor evaporates from the leaves, and more water moves up through the plant to replace the lost water.
7. Food flows down through the ____________ of the root.
8. The outermost layer of a root, stem, or leaf is called the ____________.

Answer each question.

9. Describe the path water travels through a plant. What happens to the water?

10. Name two plants whose leaves we use for food, two plants whose stems we use for food, and two plants whose roots we use for food.

Name ____________________ Date ____________

Cloze Test

Lesson 3

Roots, Stems, and Leaves

Vocabulary

secrete	minerals	food
chloroplasts	alive	water
carbon dioxide	cuticle	epidermis
photosynthesis	respiration	

Fill in the blanks.

The parts of a leaf work together to help keep the plant ____________________. The outermost layer of a leaf is its ____________________. Cells of the epidermis ____________________ a waxy coating, called a(n) ____________________. The cuticle helps keep ____________________ from leaving the leaf. The leaf makes ____________________ in cells between layers of the epidermis. These cells contain ____________________, the green food factories of plants. In addition to sunlight, these factories need water, ____________________, and the ____________________ in air to make food. Light is a form of energy that plants use to make their food. In the process called ____________________, plants trap the light to make sugar and oxygen. During the process of ____________________ plants release the energy that they use.

Name ______________________ Date __________

Lesson Outline
Lesson 4

Plant Responses and Adaptations

Fill in the blanks.  Reading Skill: **Main Idea** - questions 4, 5, 11, 13

What Are Tropisms?

1. Light, heat, and other things that produce a response are examples of a(n) ______________.
2. A plant's movement toward or away from a stimulus is called ______________.
3. Plant roots grow downward in response to ______________.
4. The response of a plant to changes in light is called ______________.
5. The roots of a willow tree show ______________ as they grow toward a water source.
6. In Darwin's experiment, the plant shoots with ______________ turned toward light, while the plants without tips did not.
7. Scientist Frits Went believed that a(n) ______________ allowed the plant shoots to bend toward light.
8. Chemicals that stimulate plant growth are called ______________.
9. Plant shoots bend toward light because auxins cause cell growth on the ______________ of the stem, making it longer.

How Do Plants Survive?

10. Adaptations help plants survive in many kinds of ______________.
11. A desert plant such as a cactus has adaptations for ______________, ______________, and ______________ water.
12. "Long-day" plants such as spinach bloom at a time of year when there is more ______________ than ______________.

Name________________________ Date____________

Lesson Outline

Lesson 4

Why Do Plants Compete?

13. Plants compete with each other for ____________, ____________, and ____________.

14. Trees in a forest have more leaves at the top in order to get more ____________.

Name ______________________ Date ______________

Interpret Illustrations

Lesson 4

What Are Tropisms?

These drawings illustrate a famous experiment in the history of science. It proved something about plant tropisms—the processes that make parts of a plant tend to turn or move in a certain direction. Darwin's experiment, pictured here, used two plant shoots to show one kind of plant tropism. Study the drawings to see what that process is.

Darwin's Experiment

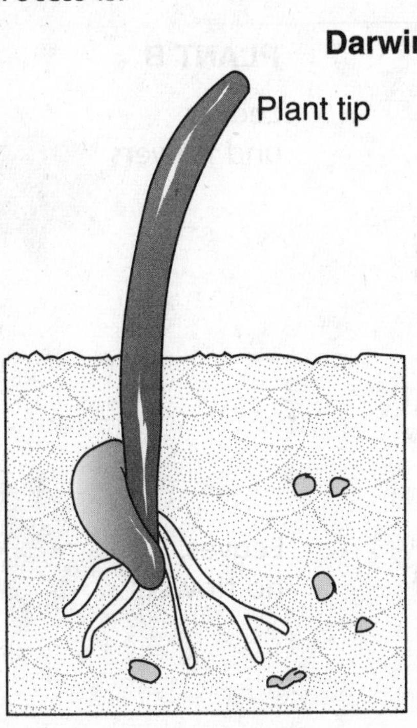

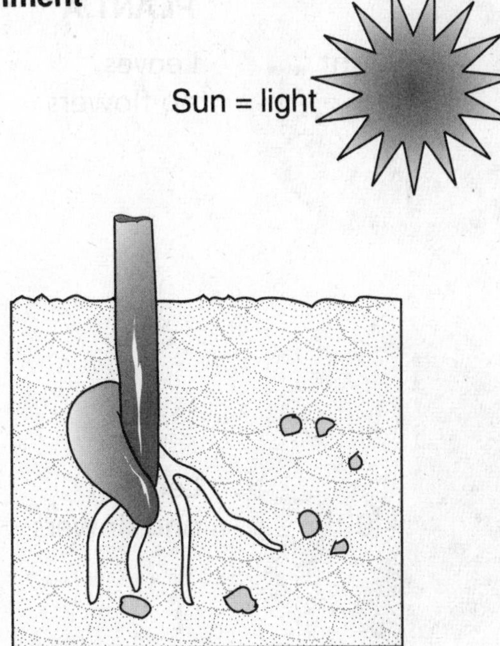

Answer these questions about the diagram above.

1. How is the plant shoot on the right different from the one on the left?

____

2. Is the plant on the right bending or growing straight up?

____

3. What part of the plant on the left is bending? ______________________

4. Toward what is the plant bending?

____

5. In your own words, summarize what this experiment showed.

____

____

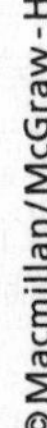

Name________________________ Date____________

Interpret Illustrations

Lesson 4

How Do Plants Survive?

The picture below shows the difference between short-day and long-day plants. In the diagram, the shaded part of the clock shows night.

SHORT DAY, LONG-DAY PLANTS

PLANT A

Leaves, no flowers

PLANT B

Leaves and flowers

Leaves and flowers

Leaves, no flowers

Answer the following questions about the picture above.

1. Which plant has flowers and leaves when there is more daylight than nighttime?

____

2. Which plant has leaves and no flowers when there is more nighttime than daylight?

____

3. Which plant's flowers would bloom during the mid-summer months?

____

4. Which plant's flowers would bloom during the fall months?

____

Name ______________________ Date ____________

Lesson Vocabulary

Lesson 4

Plant Responses and Adaptations

Vocabulary
gravitropism
short-day
stimulus
adaptation
long-day
hydrotropism
response
tropism

Fill in the blanks.

1. Anything in the environment that produces a response is called a(n) ________________.
2. If a plant bends toward a stimulus, its change is called a positive ________________.
3. A plant that grows toward the pull of gravity shows ________________.
4. Some plants, such as squash and grape plants, show a ________________ to touch.
5. A(n) ________________ helps an organism survive in its environment.
6. Plants that bloom when there is more darkness and less daylight are called ________________ plants.
7. Plants whose roots grow toward a source of water are showing ________________.
8. Plants that bloom when there is much more daylight than darkness are called ________________ plants.

Answer each question.

9. What is an adaptation? Give an example.

____

____

____

10. What are four strategies plants use to win the battle for sunlight, nutrients, and water?

____

____

____

Name ______________________ Date __________

Cloze Test
Lesson 4

Plant Responses and Adaptations

Vocabulary

storing	water	adapted
response	deserts	adaptations
night–time	center	
bloom	insects	

Fill in the blanks.

The reason plants survive in rain forests, the arctic, and ______________ is because they have ______________ to their environments. Desert plants have ______________ for collecting, ______________ and saving water. Cactus plants have a thick, waxy coating to help prevent ______________ loss. This plant's stomata open at ______________, when it's cooler. Moisture is stored in the ______________ of the cactus plant. Carnivorous plants obtain some nutrients by trapping and digesting ______________. When there is more daylight than darkness, long-day plants ______________. This flowering ______________ is known as photoperiodism.

Plant Structure and Functions

Circle the letter of the best answer.

1. The layer that separates the xylem and the phloem in a plant's stem is the
   - a. cambium.
   - b. cortex.
   - c. epidermis.
   - d. frond.

2. Some parts of plants have an outer cell called
   - a. cambium.
   - b. epidermis.
   - c. phloem.
   - d. xylem.

3. Growing roots are protected by a(n)
   - a. cortex.
   - b. epidermis.
   - c. rhizome.
   - d. root cap.

4. Almost 99% of water that enters a plant's roots is given off by
   - a. fertilization
   - b. reproduction.
   - c. respiration.
   - d. transpiration.

5. Most transpiration occurs in the
   - a. xylem.
   - b. leaves.
   - c. roots.
   - d. stem.

6. Something that causes a living thing to react is a(n)
   - a. anther.
   - b. dicot.
   - c. earned behavior.
   - d. stimulus.

7. When a plant slowly bends toward or away from a stimulus it is called a(n)
   - a. conifer.
   - b. embryo.
   - c. gravity.
   - d. tropism.

Circle the letter of the best answer.

8. A desert plant collecting, storing, and saving water is an example of a(n)

a. adaptation. **b.** reproduction.

c. competition. **d.** stimulus.

9. What a plant or animal does after receiving a stimulus is an example of a(n)

a. adaptation. **b.** answer.

c. response. **d.** sexual reproduction.

10. The tissues that carry water and minerals up through the plant are found in the

a. epidermis. **b.** fungi.

c. xylem. **d.** rootcap.

11. The layer just inside the epidermis of root and stems that stores food is the

a. cortex. **b.** xylem.

c. cambium. **d.** phloem.

12. The tissues that carry food from the leaves down through a plant are found in the

a. epidermis. **b.** aerial root.

c. xylem. **d.** phloem.

13. The threadlike parts of cells on the surface of a root are called

a. taproots. **b.** prop roots.

c. root hairs. **d.** root cap.

Name_______________________________ Date_____________

Chapter Summary

Chapter Summary

1. What are four vocabulary words you learned in the chapter? Write a sentence for each.

2. Which diagram helped you to understand an idea better?

3. What are two main ideas that you learned in this chapter?

Name________________________ Date____________

Plant Diversity

A Venn diagram is a good way to compare and contrast two things, such as how plants reproduce and how they respond. In the part of each circle that does not overlap the other, write the things that make each item different from the other. Where the circles overlap, write the things that the items have in common. Look at the example below.

Example

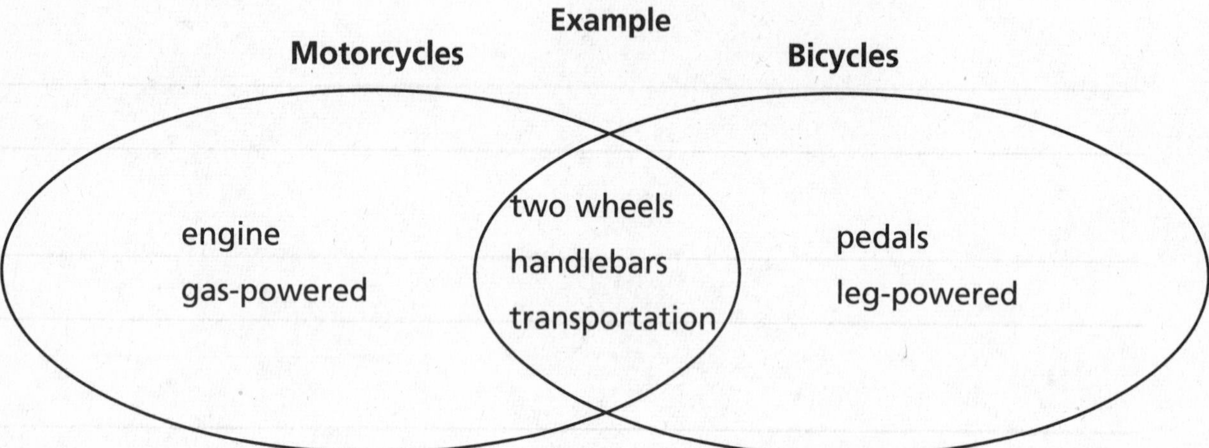

Compare and contrast angiosperms and gymnosperms by filling in the Venn diagram below. Use the words in the word bank.

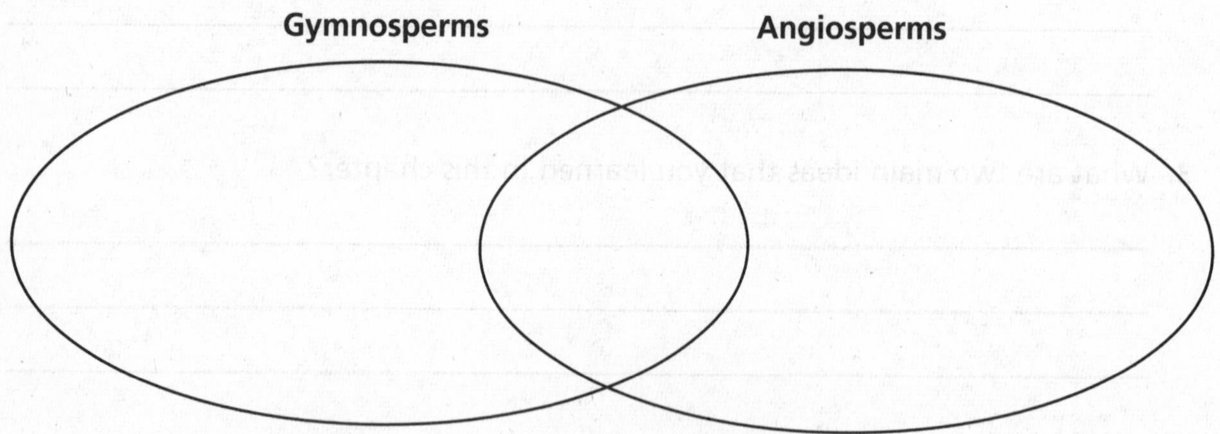

Word Bank		
cones	cotyledon	flowers
fruit	leaves	pollen
roots	seeds	stem

Use with textbook pages A54–A89

Compare and Contrast

What's the difference between **comparing** and **contrasting**? To compare things, we point out how they're alike. To contrast things, we point out how they're different.

Two things may be almost exactly alike, but not quite. Even identical twins aren't identical in all ways. One twin may smile differently or have a happier personality than the other. One may be very shy, while the other is just the opposite!

Think about each set of things below. Then write at least one way the things are alike and one way they're different.

	Alike	Different
1. A frog and a butterfly		
2. A fern and a moss		
3. A belt and suspenders		
4. A cassette tape and a CD		
5. A plane, a train, and a bicycle		
6. A book, a magazine, and a newspaper		
7. An actor, a dancer, and a painter		

Name_________________________________ Date_____________

Chapter Reading Skill

Chapter 3

Alike and Different

As you read books and articles, you may find that an author compares and contrasts people or things. To help you remember the story, make a chart on which you can compare and contrast the people or things yourself.

Read the following story, then use the chart to compare and contrast the characters. Add more features from the story to the chart.

Maria, Martina, and Mario are 12-year-old triplets. The two girls have dark hair. Their brother, Mario, is blond. They all have blue eyes.

Maria and Mario are very good singers. Martina plays piano for them. Mario also plays a mean guitar. The triplets want to start a band, but they can't seem to find the time. Mario and Martina are on the soccer team. They have practice every day after school and games on Saturdays! Maria takes dance three days a week. She often has shows or recitals on Sunday afternoons!

The kids finally had a chance to play and sing as a band. they asked a friend to play drums with them. The band performed at a celebration when Martina and Mario's soccer team won the regional championship! Maria even danced the macarena and got all the parents and guests to join in!

	Maria	Martina	Mario
Is a girl			
Is a boy			
Has dark hair			
Is blond			
Sings well			

Plants Without Seeds

Fill in the blanks.  Reading Skill: **Compare and Contrast** - questions 1, 10, 19

What Are Mosses?

1. Plants without tubelike structures to carry water are called ____________________.

2. Two types of nonvascular plants are ____________________ and ____________________.

3. Mosses are anchored in place by hairlike fibers called ____________________, which take the place of ____________________.

4. Rhizoids, like other parts of mosses and liverworts, can take in water from their ____________________.

5. While most common plants grow from seeds, mosses and liverworts grow from cells called ____________________.

6. Ferns and club mosses are vascular plants, but they are like mosses because they use spores to ____________________.

What Are Ferns?

7. The leaves, or fronds, of a fern grow from a(n) ____________________, which is an underground stem.

8. The bottom sides of some fronds are covered with ____________________.

9. Spores are spread around the fern when the spore cases ____________________.

How Do the Life Cycles of Mosses and Ferns Differ?

10. Mosses and ferns both use ____________________ to reproduce.

11. A plant that reproduces with one type of cell is reproducing by ____________________ reproduction.

12. Moss spores grow into moss plants with both ________________ and ________________ branches.

13. Male branches produce ________________, while female branches produce ________________.

14. In sexual reproduction, a male sex cell ________________ the egg, which grows into a stalk topped with a spore case.

15. Spores that are released onto damp ground grow into new ________________.

16. Both mosses and ferns alternate sexual and asexual reproduction, a process called ________________________.

17. Fern spores grow into heart-shaped plants that produce ________________ and ________________ sex cells.

18. If male and female cells from the plant join, the fertilized egg grows into a(n) ________________.

What Were the Ancestors of Plants?

19. To find the ancestors of plants, scientists first looked at other organisms that carry on ________________.

20. Green algae are like plants in various ways, including having strong ________________ containing cellulose.

21. Green algae and plants store food as ________________.

22. Scientists concluded that the first land plants to develop from algae were probably nonvascular plants similar to ________________.

Name ______________________ Date ____________

Interpret Illustrations
Lesson 5

Moss Life Cycle

The arrows indicate the direction of movement or the different stages in a process. This diagram illustrates the life cycle of mosses. Because it is a cycle—a repeating process—it does not have a beginning or an end but moves in a circle. Follow the arrows to trace the stages in this cycle.

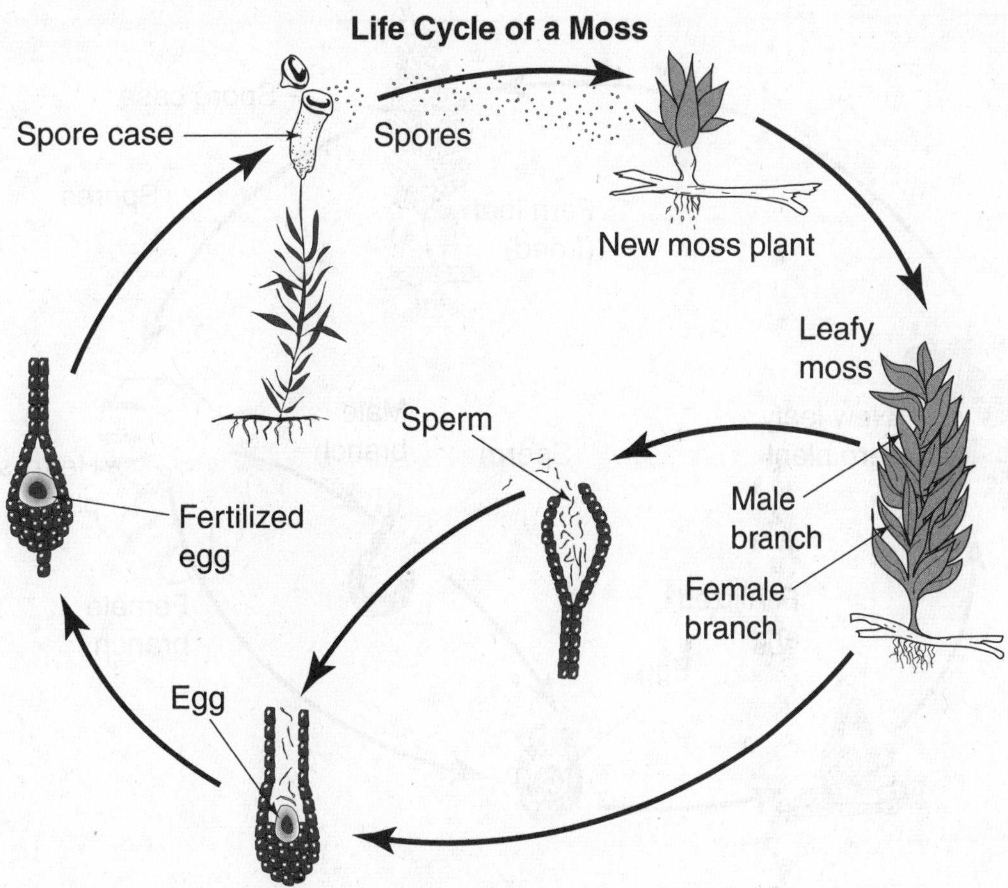

Answer these questions about the diagram above.

1. The two arrows leading from the leafy moss plant show that it produces two kinds of cells.

   a. __________________ and

   b. __________________ cells.

2. What happens when the sperm joins the egg? ______________________

3. What grows from the fertilized egg? ______________________

4. New moss plants begin to grow when __________________ are released.

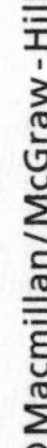

Name______________________________ Date______________

Interpret Illustrations
Lesson 5

Fern Life Cycle

In a diagram of a process or cycle, arrows are used to show the different stages in that process. This diagram shows the stages in the life cycle of a fern—how it grows and produces new fern plants. By following the arrows in the diagram, you can follow that cycle.

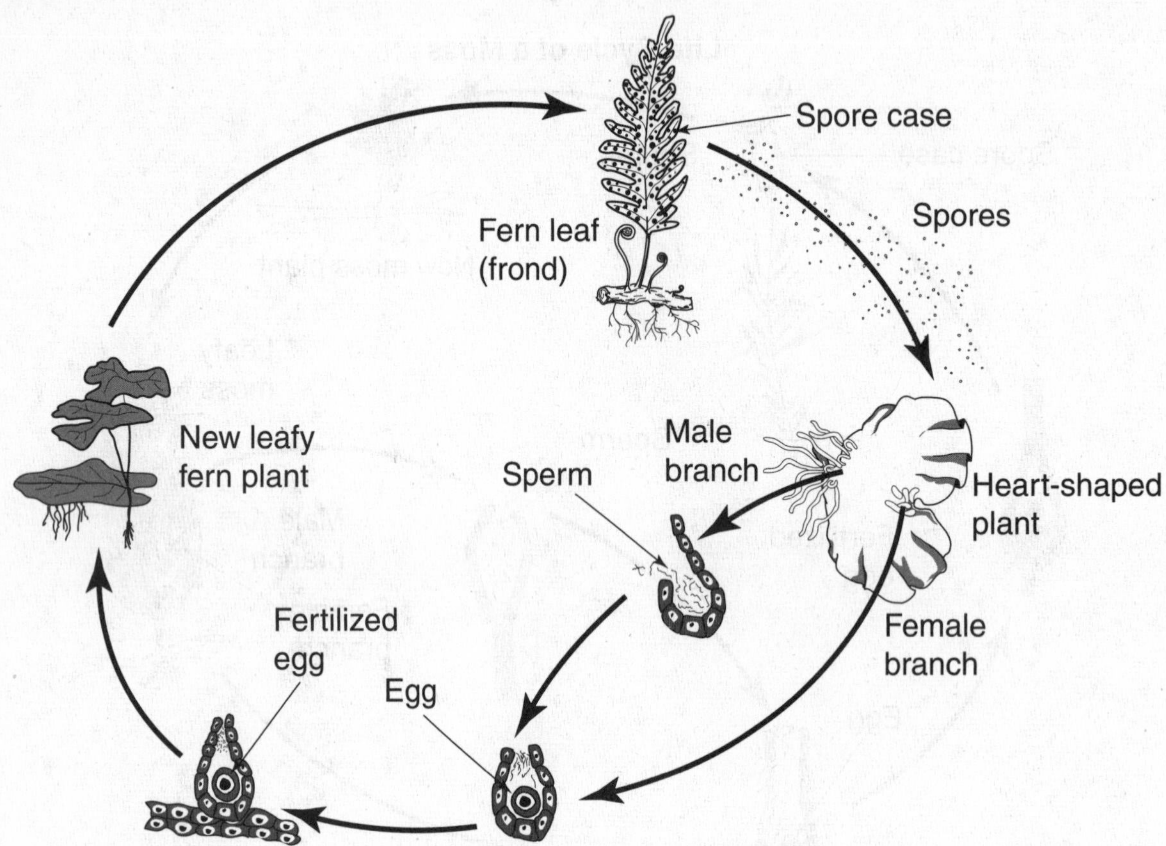

Answer these questions about the diagram above.

1. Where are the spore cases of a fern located?

____

2. After fern spores are released, what is the next stage in the cycle?

____

3. When a sperm cell and an egg cell join, the next stage in the cycle is a(n)

____

4. Does a new leafy fern plant grow from spores or from a fertilized egg?

____

Name ______________________ Date ____________

Lesson Vocabulary

Lesson 5

Plants Without Seeds

Vocabulary

vascular
sexual
fertilization
spores
fronds
rhizoids
asexual
mosses
rhizome

Fill in the blanks.

1. The hairlike fibers that hold mosses to the ground are called ________________.
2. The stage when a plant needs both male and female sex cells in order to reproduce is called ________________ reproduction.
3. Fronds grow above the ground from an underground stem called a(n) ________________.
4. The production of spores is a form of ________________ reproduction.
5. The first land plants to develop from algae were similar to ________________.
6. Ferns are ________________ plants.
7. Mosses grow from special cells called ________________.
8. The leaves of a fern are called ________________.
9. When a male sex cell and a female sex cell join together, ________________ takes place.

Answer the following questions.

10. Explain the alternation of generations.

____

____

____

11. How are mosses and ferns alike? How are they different?

____

____

____

____

Name ________________________ Date ____________

Cloze Test

Lesson 5

Plants Without Seeds

Vocabulary

liverworts	pillows	vascular
trees	spores	cells
centimeters	rhizoids	seedless

Fill in the blanks.

Neither mosses nor their close relatives the ________________, have roots. Hairlike fibers called ________________ anchor these plants in place. Both cling to sheltered rocks, moist soil, and the shady side of ________________. Mosses and liverworts are nonvascular, without the long tubelike structures that ________________ plants have. These are tiny plants, only 2 to 5 ________________ high. Mosses' leaves grow only one or two ________________ thick. Mosses resemble green, fuzzy ________________. Liverworts look like flat leaves. Both are ________________ plants, growing from ________________.

Name ______________________________ Date ____________

Lesson Outline

Lesson 6

Plants with Seeds

Fill in the blanks.  Reading Skill: **Compare and Contrast** - questions 2, 16, 17, 18

How Do Seed Plants Differ?

1. The seeds of plants contain a(n) ______________________ and ______________________.
2. Seed plants that produce flowers are ______________________, while those that do not flower are ______________________.
3. Evergreen trees, such as pines and firs, are ______________________.
4. Gymnosperms first appeared on Earth about ______________________ years ago.
5. Most fruits, nuts, and vegetables we eat come from ______________________.
6. The four divisions of gymnosperms are ______________________, ______________________, ______________________, and ______________________.
7. Gymnosperms are alike in producing their seeds on the ______________________ of female cones.
8. The leaves of most gymnosperms are shaped like ______________________.
9. While most gymnosperms are evergreen, the larch is ______________________ and loses its leaves each fall.

What Is the Life Cycle of a Conifer?

10. A pine tree produces ______________________ and ______________________ cones on a mature tree.
11. If male pollen grains happen to land on a female cone, a sperm cell from the pollen may ______________________ an egg cell in the female cone.
12. When pine cones fall from trees, their ______________________ may sprout and grow new pine trees.

Name ______________________ Date ____________

Lesson Outline

Lesson 6

What Are Angiosperms?

13. Angiosperms are the largest division in the plant kingdom, with about ____________ different members.

14. The smallest angiosperm is a flowering plant that floats on the water called ____________.

15. Gymnosperms live mainly in cold, dry, northern climates, but ____________ live in all climates and all parts of the world.

16. Angiosperms that live off other plants are ____________.

17. Two ways to distinguish an angiosperm from a gymnosperm are:

a. ______________________________, and

b. ______________________________.

What Are Cotyledons?

18. The two classes of angiosperms are based on the number of ____________, leaflike structures inside a seed, they have.

19. Besides having one cotyledon, ____________ have parallel leaf veins.

20. Maple trees and roses are dicots, which have ____________ cotyledons.

What Is the Life Cycle of an Angiosperm?

21. In an angiosperm, the female part is called the ____________ and the male part is called the ____________.

22. The transfer of pollen grains from a flower's stamen to its pistil is called ____________.

23. The fertilized cell eventually becomes a(n) ____________.

Name ________________ Date ________

What Is the Life Cycle of a Conifer?

This diagram shows the stages in the life cycle of a conifer. By following the arrows in the diagram, you can follow that cycle.

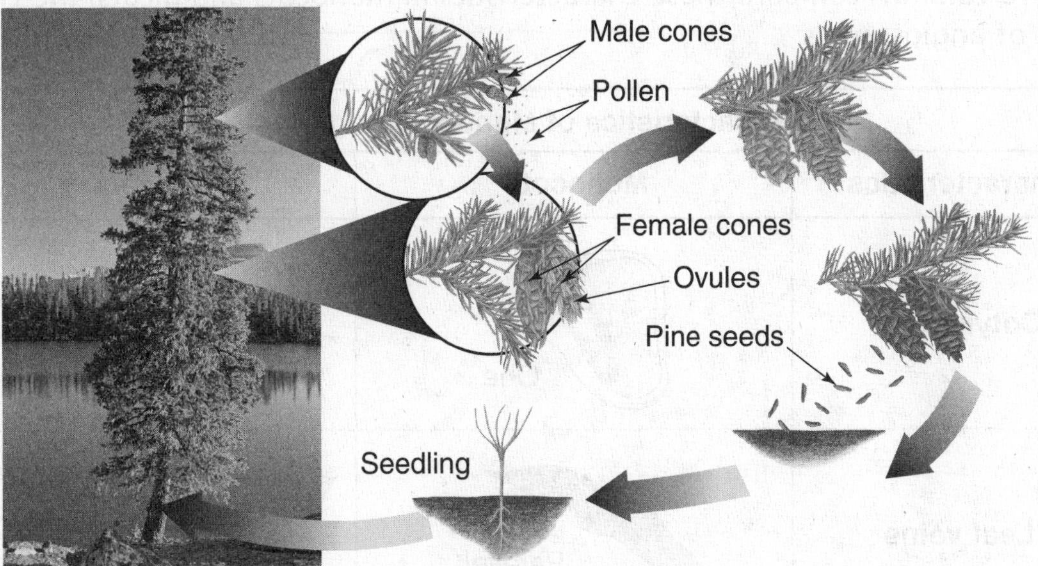

Use the diagram to answer the questions.

1. The diagram shows that a pine tree has two kinds of cones. Name the two kinds.

   ____

2. Where are the ovules located? The pollen grains?

   ____

3. In what structure do the pollen grains come in contact with the ovule?

   ____

4. Which life cycle stage occurs after pollen grains fertilize an ovule?

   ____

5. From which structure does a new pine tree grow?

   ____

6. Name the life cycle stage that occurs when a new pine tree starts growing.

   ____

Name ____________________ Date __________

What Are Cotyledons?

A chart like the one below organizes information to show similarities and differences. To read a chart, start with the title. It tells you what the overall subject is. Then read the headings for each column or section. In this chart, for example, the first column, headed "Characteristics," lists four characteristics of seed plants. The next two columns compare those characteristics in monocots and dicots, the two classes of angiosperms.

Characteristics of Monocots and Dicots

Characteristics	Monocots	Dicots
Cotyledons	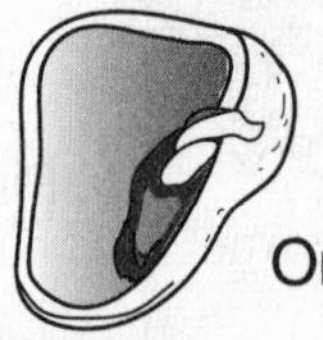 One	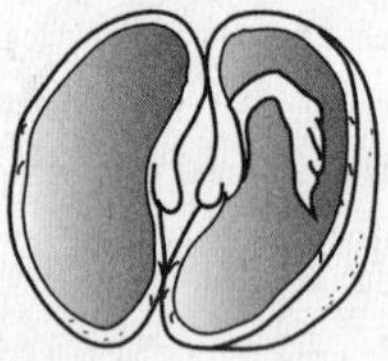 Two
Leaf veins	 Parallel	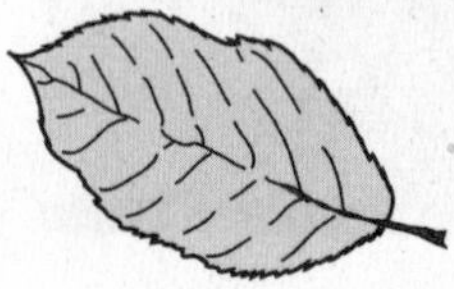 Branched
Flower parts	 Multiples of three	 Multiples of four or five
Vascular system	 Scattered in bundles	 In rings

Answer these questions about the diagram above.

1. Which class of plant—monocot or dicot—has two cotyledons?

   ____

2. The leaf veins of a dicot are ____________________.

3. The flower parts of a monocot occur in multiples of ____________________.

4. In the stem of a ____________________, the vascular system is scattered in bundles.

Name ______________________ Date ____________

Plants with Seeds

Vocabulary

- seed
- cotyledon
- gymnosperms
- dicots
- angiosperms
- pollination
- fruit
- parasite
- monocots

Fill in the blanks.

1. Angiosperms whose seeds contain two cotyledons are called ______________ for short.
2. A(n) ______________ contains an undeveloped plant and stored food.
3. Pollen grains move from a flower's stamen to its pistil during ______________.
4. As the seeds develop, the surrounding ovary enlarges and becomes the ______________.
5. Seed plants that produce flowers are called ______________.
6. Plants whose seeds contain only one cotyledon are called ______________ for short.
7. Seed plants that do not produce flowers are called ______________.
8. A tiny leaflike structure inside a seed is a(n) ______________.

Answer each question.

9. What characteristics could you use to help you decide if a plant was a monocot or a dicot?

____

____

____

____

____

10. What characteristics could you use to help you decide if a plant was a gymnosperm or an angiosperm?

____

____

Name________________________________ Date____________

Cloze Test

Lesson 6

Plants with Seeds

Vocabulary

evergreens	**conifers**	**female**
leaves	**flowers**	**Gymnosperms**
deciduous	**cuticle**	**seed**

Fill in the blanks.

__________________ first appeared when most of Earth was cold and dry, about 250 million years ago. Unlike angiosperms, gymnosperms do not produce __________________. Gymnosperms reproduce from seeds found on the scales of __________________ cones. Concealed within each __________________ is an undeveloped plant and stored food. The __________________ of gymnosperms are needle-like and are constantly being replaced. Most gymnosperms are __________________. Some conifers are called __________________ meaning they lose leaves each fall. The needles of __________________ have a small surface area and are covered with a thick __________________.

Flowers and Seeds

Fill in the blanks.  Reading Skill: **Compare and Contrast** - questions 9, 10, 12,

How Do Flowers Differ?

1. A flower that has sepals, petals, stamens, and pistils is termed a(n) ______________________.

2. A flower containing both male and female parts is called a(n) ______________________.

3. A "female" flower has pistils, but no ______________________ and would be called a(n) ______________________ flower.

4. A single oak tree has both ______________________ and ______________________ flowers, but a willow tree has either male flowers or female flowers.

5. Holly berries are the ______________________ of a holly tree with female flowers.

What Are Pollination and Fertilization?

6. Both large and tiny seeds ______________________ in the same way.

7. Pollen on a flower's ______________________ contains male sex cells.

8. The transfer of pollen from anther to stigma is called ______________________.

9. The transfer of pollen within the same flower is called ______________________, while carrying the pollen to another flower is called ______________________.

10. A pollen grain forms a tube leading to the flower's ______________________, where sperm ______________________ an egg cell.

11. A fertilized egg cell develops into a(n) ______________________.

What Is in a Seed?

12. A seed contains an immature plant, or the ______________________, and a tough covering, or the ______________________.

Name ______________________ Date ____________

13. The two steps in growing a new plant from a seed are ______________ and ______________, or sprouting.

14. To grow, seeds often must move far away to avoid ______________ from other plants.

15. Animals that eat ripe ______________ help spread seeds through their ______________.

16. Wind and animals also move the cones and seeds of ______________, which do not have fruit.

Why Do Flowers Have Aromas?

17. The main purpose of a flower's aroma is ______________.

18. Inside a flower, insects pick up grains of ______________, which contain ______________.

19. When the insect travels to another flower, it transfers pollen to ______________, helping the plant reproduce.

Name ______________________ Date ____________

Interpret Illustrations

Lesson 7

Ways Plants Are Pollinated

These diagrams show two ways in which plants are pollinated. To compare the two methods, first read the explanations under the drawings. Then read the labels on the drawings and follow the leader lines to identify each part of a flower. Notice that the flowers have similar parts.

Pollination

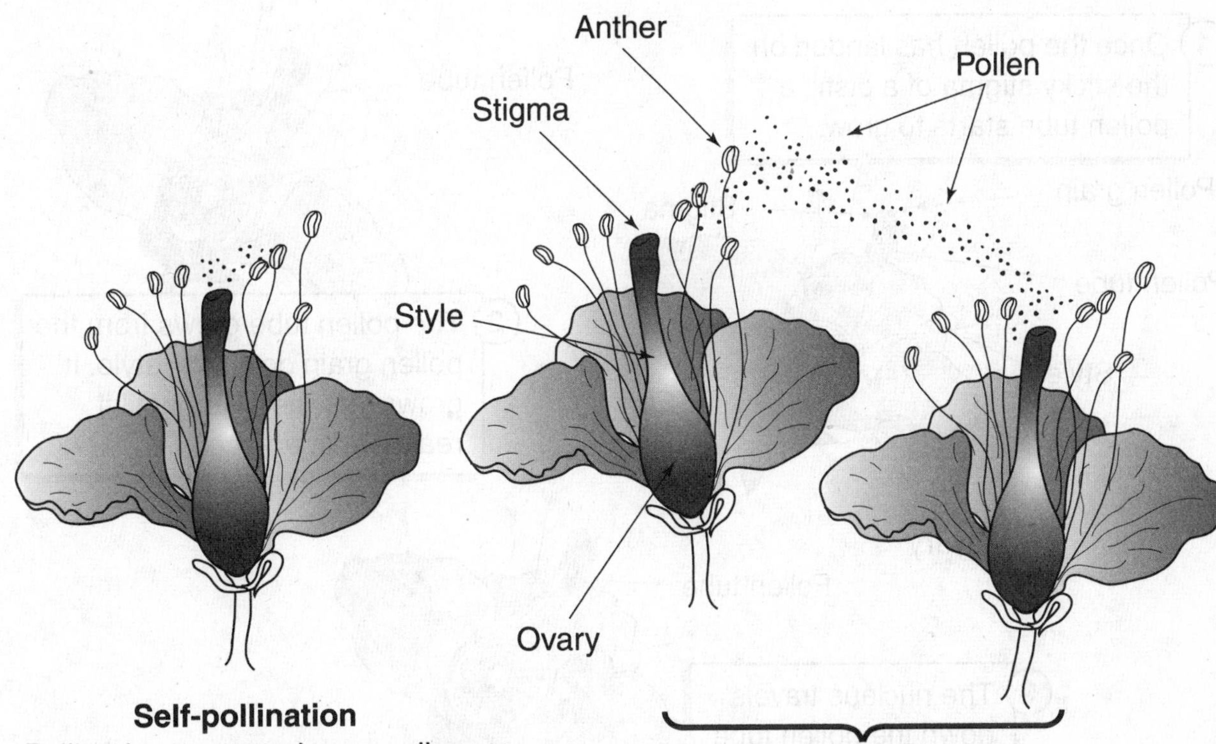

Self-pollination

Pollination occurs when a pollen grain from an anther reaches the stigma. This flower is pollinating itself because its own pollen is reaching its own stigma.

Cross-pollination

Pollination can occur between two or more flowers on separate plants. Here the pollen of one flower reaches the stigma of another.

Answer these questions about the diagram above.

1. Grains of pollen come from the ________________ of a flower.
2. The top of the flower's style is called the ________________.
3. Which method of pollination requires at least two flowers?

   ________________.
4. In both types of pollination, pollen is transferred to the ________________ of a flower.

Name________________________________ Date______________

Interpret Illustrations

Lesson 7

Process of Fertilization

These flower diagrams show the process of fertilization that takes place after pollination. Notice that each step in the process is numbered. Read the labels on the diagrams to identify each part of a flower. As you read each step, check the diagram and locate the parts that are mentioned.

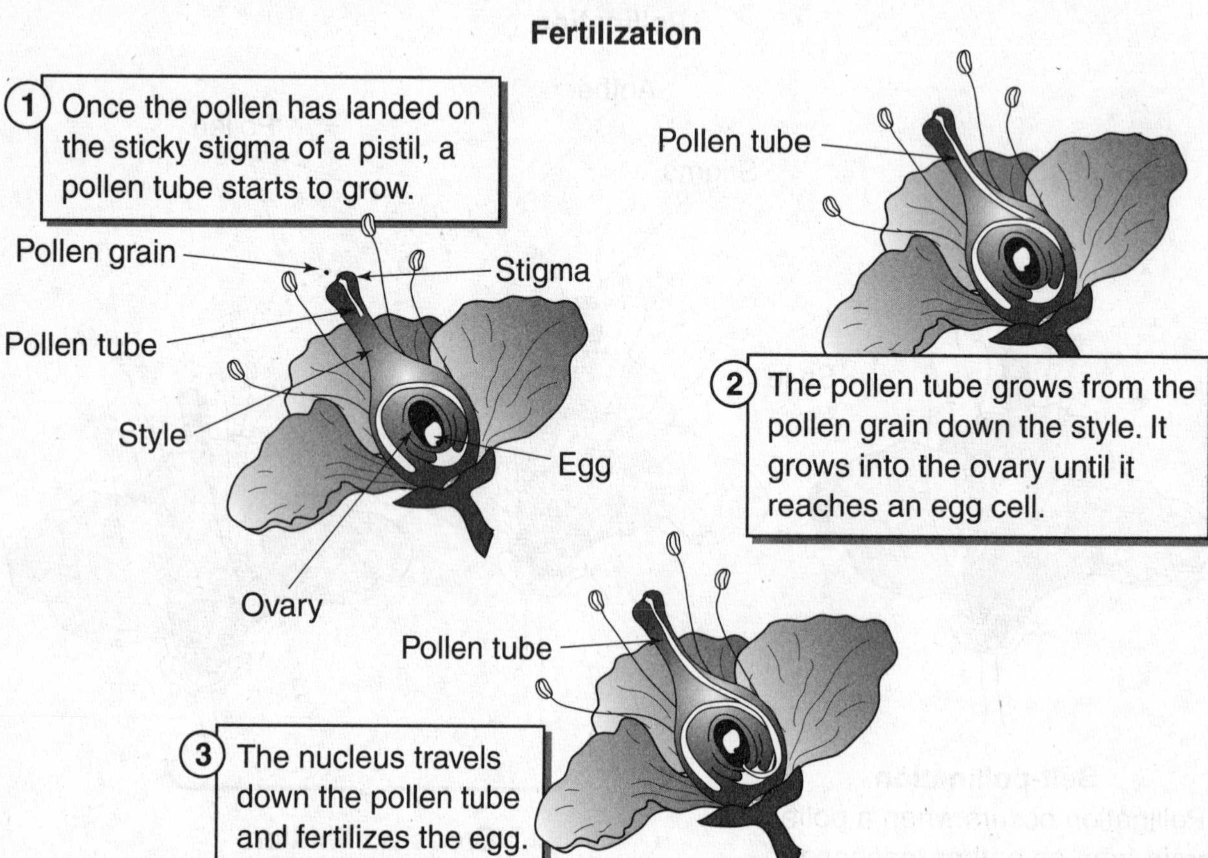

Fertilization
1 Once the pollen has landed on the sticky stigma of a pistil, a pollen tube starts to grow.
Pollen grain
Stigma
Pollen tube
Style
Egg
Ovary
Pollen tube
2 The pollen tube grows from the pollen grain down the style. It grows into the ovary until it reaches an egg cell.
Pollen tube
3 The nucleus travels down the pollen tube and fertilizes the egg.

Answer these questions about the diagram above.

1. Where do pollen grains land to cause the pollen tube to grow?

____

2. Where are the egg cells of a flower located? ____________________

3. The new structure that grows from a pollen grain is called the

____

4. How does sperm reach and fertilize the egg?

____

Name ______________________ Date ____________

Lesson Vocabulary

Lesson 7

Flowers and Seeds

Vocabulary

fruit
pistil
pollination
ovary
embryo
seed coat
anther
wind

Fill in the blanks.

1. The transfer of pollen from the anther to the stigma is called ________________.
2. A(n) ________________ is an immature plant inside a seed.
3. A mature ripened ovary of a plant is called the ________________.
4. Many seeds are scattered by the ________________.
5. Grains of pollen are found in the ________________ of a flower.
6. The base of the pistil is the ________________.
7. The ovary in a flower is found at the base of the ________________.
8. The outer covering of a seed is called a(n) ________________.

Answer the following questions.

9. Describe the life cycle of a conifer.

____

____

____

____

10. What is the difference between self-pollination and cross-pollination?

____

____

____

____

Name______________________________ Date______________

Cloze Test

Lesson 7

Flowers and Seeds

Vocabulary

dispersal	competition	warm	germination
wastes	ovary	wind	

Fill in the blanks.

Before seeds produce plants, they move to where they can sprout. This is known as seed ____________________. Seeds travel far from the parent plant to avoid ____________________ for sunlight and water. Some seeds have parachutes that are blown by the ____________________. Others are transported by animals that eat the plant's ripened ____________________, which is called a fruit. Animals deposit the seeds in their ____________________. For sprouting, which is called ____________________, seeds need water and a(n) ____________________ temperature.

Name ______________________ Date ____________

Chapter Vocabulary

Chapter 3

Plant Diversity

Circle the letter of the best answer.

1. Plants that produce flowers are
   a. angiosperms. b. conifers.
   c. gymnosperms. d. tropisms.

2. Monocots contain only one
   a. conifer. b. cotyledon.
   c. gymnosperm. d. response.

3. A fruit is a seed's ripened
   a. anther. b. gymnosperm.
   c. ovary. d. stigma.

4. Most gymnosperms are
   a. angiosperms. b. evergreens.
   c. deciduous. d. violets.

5. Conifers that lose their leaves in the fall are
   a. evergreens. b. deciduous.
   c. saplings. d. young.

6. Cross-pollination occurs when
   a. the pollen of one flower reaches the stigma of another.
   b. a grain of pollen from an anther reaches the stigma.
   c. the pollen is transferred from the stigma of one flower to the anther of another.
   d. a grain of pollen is transferred from the stigma of one flower to the anther of another.

7. Angiosperms are divided into two classes by their number of
   a. taproots. b. dicots.
   c. cotelydons. d. monocots.

Circle the letter of the best answer.

8. Instead of roots, mosses and liverworts have
   a. spores.
   b. fronds.
   c. rhysomes.
   d. rhizoids.

9. Ferns have leaves that are called
   a. rhysomes.
   b. angiosperms.
   c. fronds.
   d. rhizoids.

10. Some plants like mosses and liverworts do not grow from seeds but from
   a. taproots.
   b. rhysomes.
   c. ferns.
   d. spores.

11. The stage in life cycle of mosses that requires only one type of cell to reproduce is called
   a. cross pollination.
   b. fertilization.
   c. asexual reproduction.
   d. sexual reproduction.

12. The stage in the life cycle of mosses that requires both male and female sex cells to reproduce is called
   a. sexual reproduction.
   b. self-pollination.
   c. asexual reproduction.
   d. fertilization.

13. The joining of male sex cell and a female sex cell is called
   a. sexual reproduction.
   b. cross pollination.
   c. asexual reproduction.
   d. fertilization.

Chapter Summary

1. What are four vocabulary words you learned in the chapter? Write a definition for each.

2. Which diagrams explain an idea the best?

3. What are two main ideas that you learned in this chapter?

Name________________________ Date______________

Chapter Graphic Organizer
Chapter 4

Animal Diversity

A graphic organizer like the one shown below can help you summarize a paragraph. Choose a paragraph from your textbook. Fill in the details for your paragraph under the heading marked "Details." Then, write your summary under the heading marked "Summary". You can see if your summary is correct by checking it against the details that you listed. Each detail should support your summary.

Details
1.
2.
3.
4.
5.
6.
7.
8.

Summary

Summarize

A summary should reflect the main idea of a paragraph and be supported by each of the details in the paragraph. Read the following paragraph entitled "Blending In." Underline each of the details in the paragraph. Then write a summary in the space provided.

Blending In

About 150 years ago, England was home to two kinds of peppered moths. One kind was light colored. The other kind was dark colored. Birds fed on both kinds of moths, many of which clung to the trunks of trees. However, gradually the light-colored moths seemed to be disappearing. The birds were eating more of these moths than the dark-colored ones. This happened because nearby factories were pouring dark, sticky smoke into the air. The smoke stuck to the trunks of trees. The light-colored moths stood out against this background. The dark-colored moths blended in with the background. Since the birds could more easily see the light-colored moths, the birds were eating more light-colored moths than dark-colored moths. This is an example of a kind of camouflage called protective coloration. The color of the dark peppered moths protected them from predators.

Summary

More Summarizing Paragraphs

A summary should also be short and easy to understand. Read the following paragraph entitled "Cells, Tissues, Organs, and Systems." Then choose the summary from the ones listed below that best reflects the paragraph.

Cells, Tissues, Organs, and Systems

The building blocks of living things are cells. Whether an animal grows hair or feathers depends on the kinds of cells it has. Scientists often study the cells of animals in order to best group them. Similar cells that have the same job or function come together to make a tissue. Tissues of different kinds come together to make an organ, like a heart or brain. Finally, a group of organs that work together to do a certain job make up an organ system. For example, an animal's digestive system includes its mouth, stomach, and intestines.

Summary

1. The building blocks of living things are cells.
2. Scientists often study the cells, tissues, and organs of animals in order to best group them.
3. Cells, the building blocks of all living things, form tissues; tissues come together to make organs; organs come together to make organ systems.
4. A heart is an organ.

Name ______________________ Date ____________

Lesson Outline

Lesson 8

Animal Traits

Fill in the blanks.  Reading Skill: **Compare and Contrast** - 3

What Are the Traits of Animals?

1. Animals are many-celled ________________ that are made of different kinds of cells.
2. Organ ________________ enable animals to perform different functions.
3. The most diverse vertebrate group is the ________________ group.
4. Humans belong to a class of vertebrates called ________________.
5. An organ ________________ is a group of organs that work together to do a certain job.
6. Animals that have backbones are members of a group called ________________.
7. Members of the group of animals called ________________ have no backbones.
8. One of the simplest kinds of animals is a sponge which belongs to a group called ________________.
9. Hydra, anemones, and jellyfish belong to a group called ________________ and do not have heads or tails.

10. ________________, or flatworms, have flat bodies but no true organ systems.

11. Segmented worms or ________________ have specialized organs such as jaws or gills as well as a circulatory, digestive, and nervous system.

12. The ________________ group, or mollusks, include snails, clams, and octopuses and have a shell which may be either inside or outside of the mollusk's body.

13. Members of the largest animal group, ________________, have tough outer skeletons, jointed legs, and a body made up of two, three, or more sections.

14. Unlike the arthropods, the ________________ or echinoderms, have a skeleton inside their bodies.

15. Sharks and rays, which are part of the ________________ group, have cartilage running down their backs in a chain of smaller parts.

16. Frogs, toads, and salamanders are ________________ that live in water during the early stages of life.

17. The ________________, the first group of vertebrates to grow and develop out of the water, breathe through their lungs and have waterpoof scales on their skins.

18. Members of the ________________ group have feathers, wings, and strong but lightweight bones.

Name ____________________ Date ____________

Interpret Illustrations
Lesson 8

What Are the Traits of Animals?

The illustrations below show two very different animals. Animals are classified into groups by their traits. Animals with similar traits are classified into the same groups.

Use the illustrations to answer the questions.

1. What are two characteristics of both the octopus and the frog that make them animals?

____

____

2. Describe three ways in which these animals are the same.

____

____

3. Describe three ways in which these animals are different.

____

____

____

4. Do you think the frog and the octopus should be in the same group? Why or why not?

____

____

Name ______________________ Date ____________

Interpret Illustrations
Lesson 8

What Are Invertebrates?

The animal kingdom is divided into two main groups. One group called vertebrates is made up of animals that have backbones. The other group called invertebrates is made up of animals that do not have backbones. The picture below shows some of the types invertebrates.

Starfish

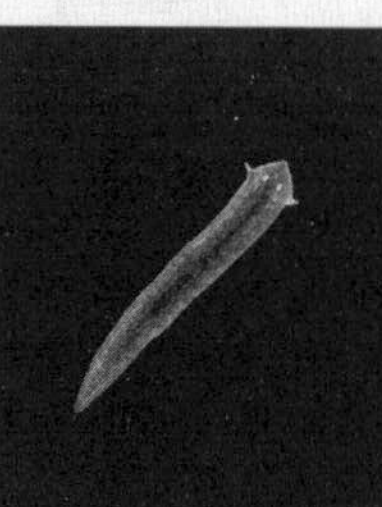
Planarian

Sponge

Octopus

Grasshopper

Tree snail

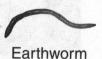

Earthworm

Sea anemone

Hydra

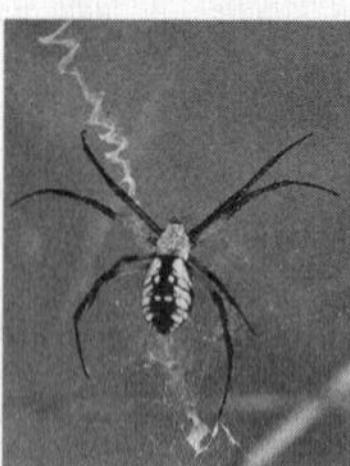
Spider

Marine flatworm

Lobster

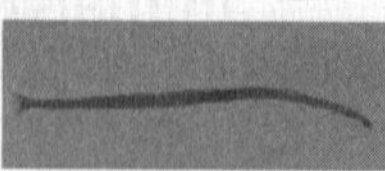
Roundworms

Answer the following questions about the picture above.

1. What is the main difference between vertebrates and invertebrates?

____

____

2. What group do the octopus and the tree snail belong to? What three main body parts do they have?

____

____

3. Name three groups of invertebrates that live in water.

____

____

Name ______________________________ Date ____________

Lesson Vocabulary

Lesson 8

Animal Traits

Vocabulary

- amphibian
- vertebrates
- mammal
- organ system
- fish
- invertebrates
- reptile
- bird

Fill in the blanks.

1. Animals that have backbones are called ____________________.

2. Animals that do not have backbones are called ____________________.

3. A(n) ____________________ enables an animal to sense its environment, get rid of wastes, and reproduce.

4. Vertebrates are divided into five classes:

   ____

   ____

   ____

   ____

   ____

Answer each question.

5. What is the major difference between vertebrates and invertebrates?

   ____

   ____

   ____

6. What structure does bony fish, or Osteichthys, have that sharks do not have? How does this structure help bony fish to function?

   ____

   ____

   ____

Name ______________________________ Date ______________

Cloze Test

Lesson 8

Animal Traits

Vocabulary

gills	water	moist	vertebrates	adapted
fish	reptiles	scales	invertebrates	groups

Fill in the blanks.

Animals are divided into two large groups. One group called ____________________ is made up of animals that have backbones. The members of the other group have no backbones. They are called ____________________. Vertebrates and invertebrates can be divided into smaller ____________________. The most diverse vertebrate group is ____________________. Most fish have ____________________ and are usually covered by scales. Amphibians are ____________________ to live part of their lives in ____________________ and part of their lives on land. Amphibians have thin, ____________________ skin. Snakes, turtles, alligators, and lizards are ____________________. They have dry skin covered with ____________________.

Name ______________________________ Date ______________

Lesson Outline
Lesson 9

Animal Adaptations

Fill in the blanks.  Reading Skill: **Summarize** - questions 1, 2, 3, 5, 6, 7, 8, 9, 18

How Do Animals Adapt?

1. An important group of adaptations helps an animal keep from getting eaten by a(n) ______________.

2. In nature, looking like something else, especially something unpleasant, is called ______________.

3. Monarch ______________ (caterpillars) are protected from predators because they feed on milkweed which contains a substance that can make many animals ill.

4. ______________ look like thorns, so predators stay away from them.

What Is Camouflage?

5. An animal that does not move, or moves very, very slowly and looks like its surroundings is using ______________ to protect itself.

6. The wings of a leaf ______________ are shaped like the leaves of a plant which makes it very hard to see.

7. Dark-colored moths that blend in with the background is an example of ______________.

What Is Inherited?

8. A(n) ______________ behavior is one that is not learned and is passed down from one generation to the next.

9. Inherited behavior, which is done automatically, is called ______________.

10. The passing of inherited traits from parents to offspring is called ____________________.
11. Some traits are affected by both heredity and the ____________________.

What Is a Hybrid?

12. Living things that have parents that are quite different from each other, such as donkeys and horses, are called ____________________.
13. People sometimes ____________________ hybrids on purpose, since a hybrid may have more desirable traits than either of its parents.
14. Some people mate living things that are closely related in a process called ____________________.
15. A ____________________ is a product of the mating of individuals from two distinct breeds or varieties of the same species.

Why Is Diversity Important?

16. Animal ____________________ refers to a group of the same kind of animal in which there are lots of animals with different traits.
17. When an environment changes, only those animals that can ____________________ to the change will survive.
18. If the population is made up of animals with the same traits, and those traits do not help the animals survive in a changing environment, the whole ____________________ may die out.

Name ______________________ Date ____________

Interpret Illustrations

Lesson 9

How Do Animals Adapt?

The picture below shows two similar-looking insects. One is a yellow jacket which has a poison stinger. The other is a harmless fly called a syrphid. The syrphid is using an adaptation called mimicry to protect itself from predetors.

Answer the following questions about the picture above.

1. How do the yellow jacket and the syrphid look similar?

____

____

2. How do the yellow jacket and the syrphid look different?

____

____

3. How does looking like a yellow jacket help the syrphid protect itself from predetors?

____

____

Name ______________________ Date ______________

Interpret Illustrations

Lesson 9

What Is Camouflage?

The pictures below show two kinds of peppered moths against a dark tree. The darker peppered moth is harder for predators to see because it blends in with the color of the dark tree.

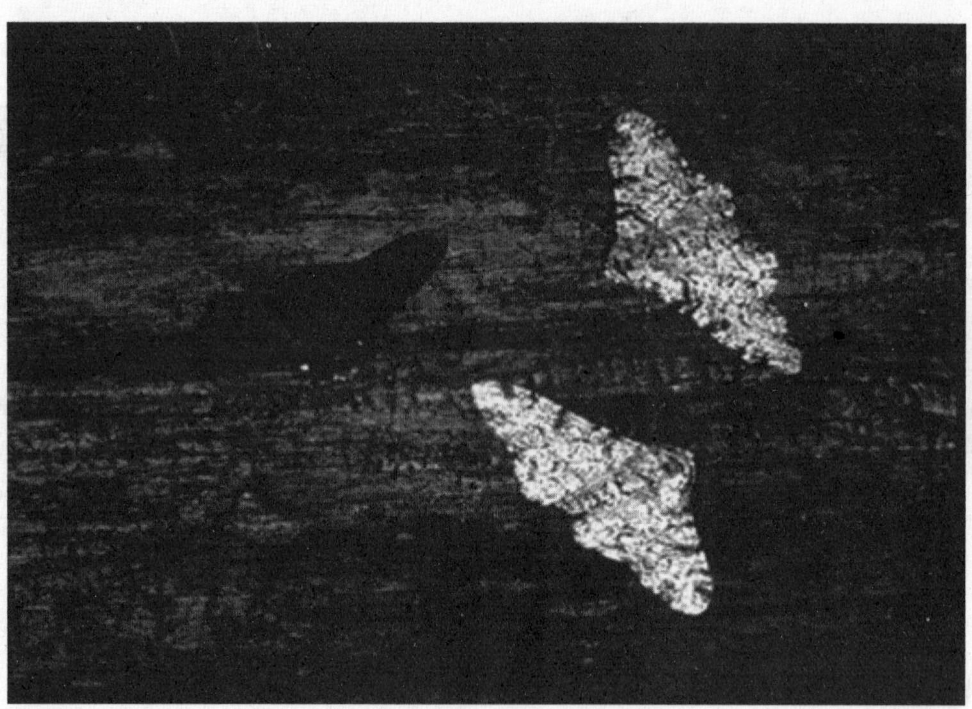

Answer the following questions about the picture above.

1. Which of the peppered moths is easier to see?

____

____

2. Why would predators eat more of the lighter colored peppered moths?

____

____

3. What would happen if the trees were much lighter in color? Why?

____

____

Name ______________________ Date ____________

Lesson Vocabulary
Lesson 9

Animal Adaptations

Fill in the blanks.

Vocabulary
mimicry
heredity
diversity
coloration
crossbreeding
instinct
hybrids
camouflage

1. An animal that does not move, or moves very, very slowly, and looks like its surroundings is using ____________ to protect itself.
2. The passing of inherited traits from parents to offspring is called ____________.
3. Living things that have parents that are quite different from each other, such as donkeys and horses, are called ____________.
4. Animal ____________ refers to a group of the same kind of animal in which there are lots of animals with different traits.
5. Looking like something else is an example of ____________.
6. Dark-colored moths that blend in with the background are an example of protective ____________.
7. Mating closely related living things is a process called ____________.
8. An inherited behavior that is done automatically is called ____________.

Answer each question.

9. Explain why animals avoid a syrphid insect even though it doesn't sting. What kind of adaptation is this protection called?

____

____

10. How is a leaf butterfly protected from being eaten? What is this kind of protection called?

____

____

____

Name ________________________ Date ______________

Cloze Test

Lesson 9

Animal Adaptations

Vocabulary

background	**shape**	**wings**
camouflaged	**protective**	**blending**
adaptation	**color**	

Fill in the blanks.

An animal that does not move, or moves very, very slowly, and looks like its surroundings is ________________. Camouflage is another important ________________ that lets animals avoid their predators. There are two basic kinds of camouflage, or ________________ in with the environment. One has to do with an animal's ________________. The other has to do with its ________________. The ________________ of the leaf butterfly are shaped like the leaves of a plant. Dark-colored peppered moths blend in with the ________________. This is an example of ________________ coloration.

Name____________________ Date__________

Animal Diversity

Circle the letter of the best answer.

1. Living things that have parents that are different species are called

   a. donkeys. b. hybrids.
   c. offspring. d. mules.

2. Animals that blend in with their backgrounds are an example of protective

   a. hiding. b. camouflaging.
   c. coloration. d. mimicry.

3. A trait that is passed on from one generation to the next is

   a. inherited. b. mimicry.
   c. diversity. d. instinct.

4. An animal that looks like another in an unpleasant way is using

   a. protective coloration. b. camouflage.
   c. mimicry. d. heredity.

5. The passing of inherited traits from parents to offspring is called

   a. inherited. b. heredity.
   c. coloration. d. crossbreeding.

6. Mating closely related living things is a process called

   a. mimicry. b. inherited traits.
   c. coloration. d. crossbreeding.

7. A group of the same kind of animal in which there are lots of animals with different traits is called animal
   a. crossbreed. b. adaptation.
   c. mimicry. d. diversity.

8. Animals that live their whole lives in water are
   a. fish. b. amphibians.
   c. reptiles. d. mammals.

9. Animals that are adapted to live part of their life in water and part of their life on land are
   a. invertebrates b. mammals.
   c. birds. d. amphibians.

10. An animal is camouflaged when
    a. it looks like another animal.
    b. it looks different from its background.
    c. it moves very, very slowly, and it looks like its surroundings.
    d. it looks like a thornbug.

11. Mammals are animals that
    a. feed their young milk. b. lay eggs.
    c. that have water proof scales d. have no backbones.

12. Snakes, alligators, turtles, and lizards are
    a. reptiles. b. invertebrates.
    c. amphibians. d. fish.

13. Vertebrates with hollow bones and air sacs that help them to fly are
    a. amphibians. b. fish.
    c. birds. d. reptiles.

Name________________________________ Date____________

Meanings and Words

Read each meaning. Then find a word in the Word Box that fits that meaning, and write it on the line.

Word Box				
tropism	stimulus	system	heredity	camouflage
cambium	vertebrates	cotyledon	mimicry	chlorophyll
frond	vascular	organ	rhizoid	conifer

Meanings **Words**

1. looking like something else ____________________
2. a green chemical in plants ____________________
3. separates the xylem from the phloem ____________________
4. blending in with the environment ____________________
5. anchors moss to the soil ____________________
6. animals that have backbones ____________________
7. the leaf of a fern ____________________
8. tissues of different kinds come together to make this ____________________
9. a tiny leaflike structure inside a seed ____________________
10. a group of organs that work together to do a certain job ____________________
11. something that makes a living thing react ____________________
12. a response to a stimulus ____________________
13. the passing of inherited traits from parents to offspring ____________________
14. composed of or having vessels ____________________
15. a tree that produces seeds in cones ____________________

Word Webs

A word web lists words that describe or relate to the same thing. Here's an example using the word *bacteria*.

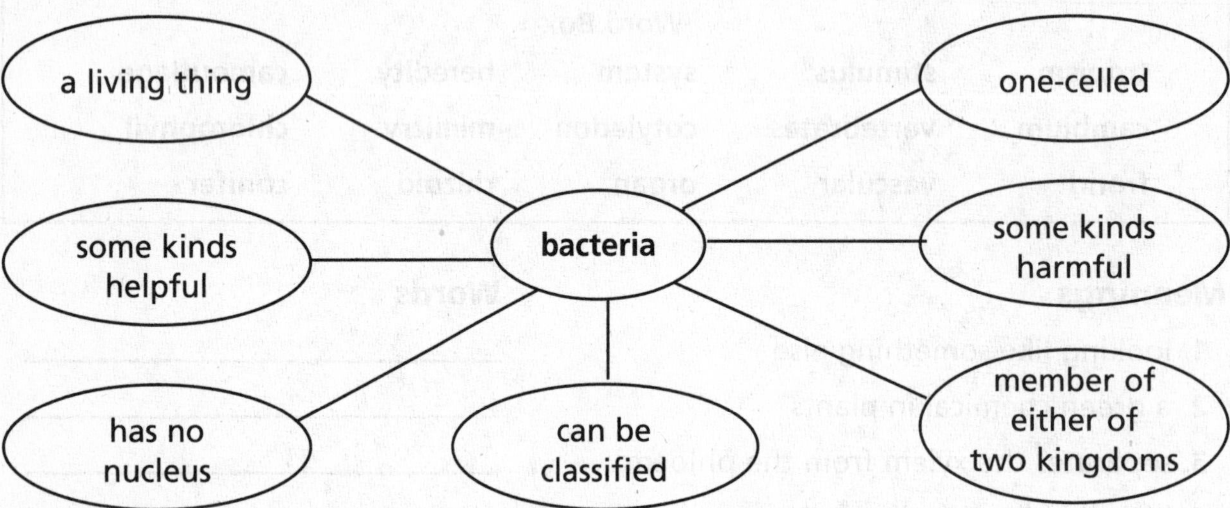

Now it's your turn. choose a vocabulary word, such as *conifer*, *tissue*, or *photosynthesis*. Make a web using other words that relate to or describe it.

Name ______________________________ Date ____________

Find-a-Word

Look across, down, and diagonally to find these hidden words:

ASK, BACTERIUM, CHLOROPHYLL, CLASSIFY, CORTEX, OVARY, DICOT, FRONDS, FRUIT, HYPOTHESIS, MOSS, NUCLEUS, PHLOEM, PHOTOSYNTHESIS, PHYLUM, PLANTS, POLLEN, PROTIST, RESPIRATION, RHIZOID, RHIZOME, ROOTS, SEEDS, SPORE, STIMULUS, STOMATA, TREE, VASCULAR, XYLEM.

P	P	O	L	L	E	N	Z	F	R	O	N	D	S	X
R	H	I	Z	O	M	E	J	X	F	R	U	I	T	Z
H	L	O	V	A	S	C	U	L	A	R	C	Z	O	J
I	O	X	T	O	V	A	R	Y	X	Y	L	E	M	C
Z	E	P	R	O	T	I	S	T	T	R	E	E	A	H
O	M	X	Z	J	S	T	I	M	U	L	U	S	T	L
I	V	J	K	P	H	Y	L	U	M	X	S	Z	A	O
D	A	S	Z	X	J	R	N	C	O	R	T	E	X	R
X	R	S	P	H	Y	P	O	T	H	E	S	I	S	O
Z	I	Z	E	O	Z	X	J	O	H	M	O	S	S	P
B	A	C	T	E	R	I	U	M	T	E	A	S	K	H
J	B	X	Z	J	D	E	C	L	A	S	S	I	F	Y
P	L	A	N	T	S	S	D	I	C	O	T	I	X	L
R	E	S	P	I	R	A	T	I	O	N	X	Z	S	L

Name ____________________ Date ____________

Chapter Graphic Organizer
Chapter 5

Interactions of Living Things

A flowchart is a good way to show a sequence of events. The events listed below are the sequence of events in a food chain. Read the first event in the first box. Then follow the arrows to the next event in the second box, and so on until you have read all the events.

Events in a Food Chain

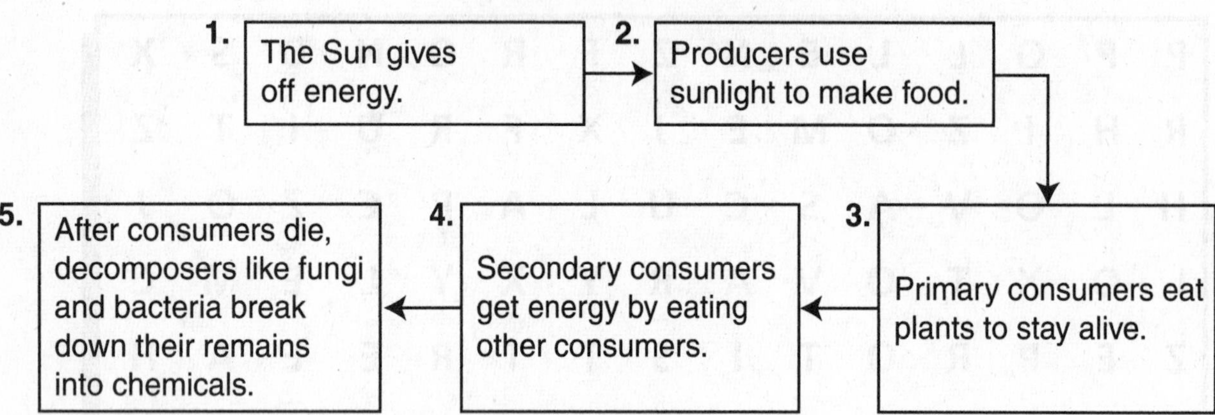

1. The Sun gives off energy.
2. Producers use sunlight to make food.
3. Primary consumers eat plants to stay alive.
4. Secondary consumers get energy by eating other consumers.
5. After consumers die, decomposers like fungi and bacteria break down their remains into chemicals.

Numerous species of plants and animals live on the Blackland Prairie. Make a flowchart of a food chain for the prairie. Fill in the blank boxes for the flowchart below.

Food Chain in the Blackland Prairie

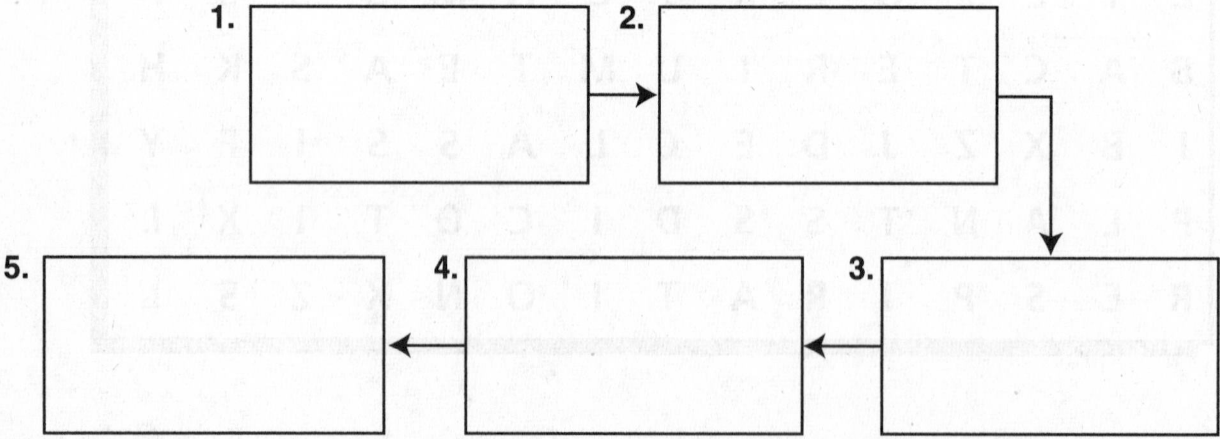

Name ______________________ Date ____________

Sequence of Events

A sequence of events tells you the order of a story. By following the order of events in a story, you will have a better understanding of the whole story.

Read the following text based on the information in your textbook. Then answer the questions below.

What Is a Food Chain?

A food chain is the path energy takes from producers to consumers to decomposers. For example, on the prairie the first organisms in a food chain are plants. Plants capture the Sun's energy during photosynthesis. The plants provide food or energy for plant eating animals such as grasshoppers. A lizard may snap up a grasshopper and get some of its stored energy. The red-tailed hawk eats snakes, mice, lizards, rabbits, and other birds. The red-tailed hawk doesn't eat plants. However, it still gets some of the Sun's energy that was originally stored in plants. The energy traveled from plant to grasshopper to lizard to hawk.

1. What is the first step in a food chain?

____

____

2. If you were to make a flowchart, what number box would you put a lizard eating a grasshopper?

____

____

3. What are the first organisms in the food chain on the prairie?

____

____

4. What is the last organism in the food chain mentioned in the above paragraph?

____

____

Put the Events in Order

Every story has a sequence. The events below are based on material in your textbook. Put each event in the correct order.

_____ Much of our food is raised or grown in the prairies.

_____ Animals that fed on the buffalo leave the Blackland Prairie.

_____ European settlers arrive from the east.

_____ Native Americans hunt buffalo for food and clothing as a means of survival.

_____ People discover that the prairie has rich topsoil.

_____ Huge herds of buffalo graze on the Blackland Prairie.

_____ Farmers start growing crops with shallow roots, such as corn, wheat, cotton, and sorghum.

_____ Buffaloes leave the Blackland Prairie.

Write a step-by-step sequence to explain why the American bald eagle was put on the endangered species list.

____

____

____

____

____

Name______________________________ Date____________

Interactions in an Ecosystem

Fill in the blanks.  Reading Skill: **Sequence of Events** - questions 10, 11, 12

What Is an Ecosystem?

1. All the interacting living and nonliving things in an area describes a(n) ____________________.
2. A nonliving factor in an ecosystem is called a(n) ____________________ factor.
3. Abiotic factors in an ecosystem include light, water, ____________________, temperature, air, and minerals.
4. The living parts of an ecosystem are called ____________________ factors.
5. Living organisms that produce oxygen and food that animals need are called ____________________.
6. Animals, or ____________________, produce the carbon dioxide that plants need.
7. Decomposers break down dead organisms and produce ____________________ that enrich the soil.

What Is a Prairie Ecosystem Like?

8. Long ago, the prairie was a "sea of wild ____________________."
9. Native Americans once hunted ____________________ on prairie lands. Today cattle and crops live there.

What Is the Treasure of the Blackland Prairie?

10. Prairie soils can often be identified by their rich, dark ____________________, or top layer of soil.
11. Partly decayed plant matter that enriches the prairie soil is called ____________________.

Name ____________________ Date __________

Lesson Outline
Lesson 1

What Animals Live on the Blackland Prairie?

12. About ______________ species of animals live on the Blackland Prairie.

13. Pipits, longspurs, horned larks, and about 300 other kinds of ______________ live on the Blackland Prairie.

What Are Populations and Communities?

14. The populations in a community ______________ with each other in different ways.

15. A person who investigates the activities of living things in an ecosystem is called a(n) ______________.

What Are Niches and Habitats?

16. The place where an organism lives is called its ______________.

17. What a species eats and what eats that species is part of its ______________, or role in the ecosystem.

How Do Organisms Change Their Environment?

18. Beavers build ______________ that back up water into ponds.

19. Beaver ponds become ______________ areas which allow certain trees to grow.

How Do Organisms Survive in Variable Environments?

20. Organisms survive difficult times by adapting to changes in their ______________.

21. In a drought, the eastern spadefoot toad covers itself with ______________ to get water.

Name ______________________________ Date ______________

Interpret Illustrations

Lesson 1

What Is an Ecosystem?

This diagram shows two different pictures of what could be a single environment. One picture focuses on abiotic factors and the other on biotic factors. The small drawings in each picture show close-ups of some of the smaller factors.

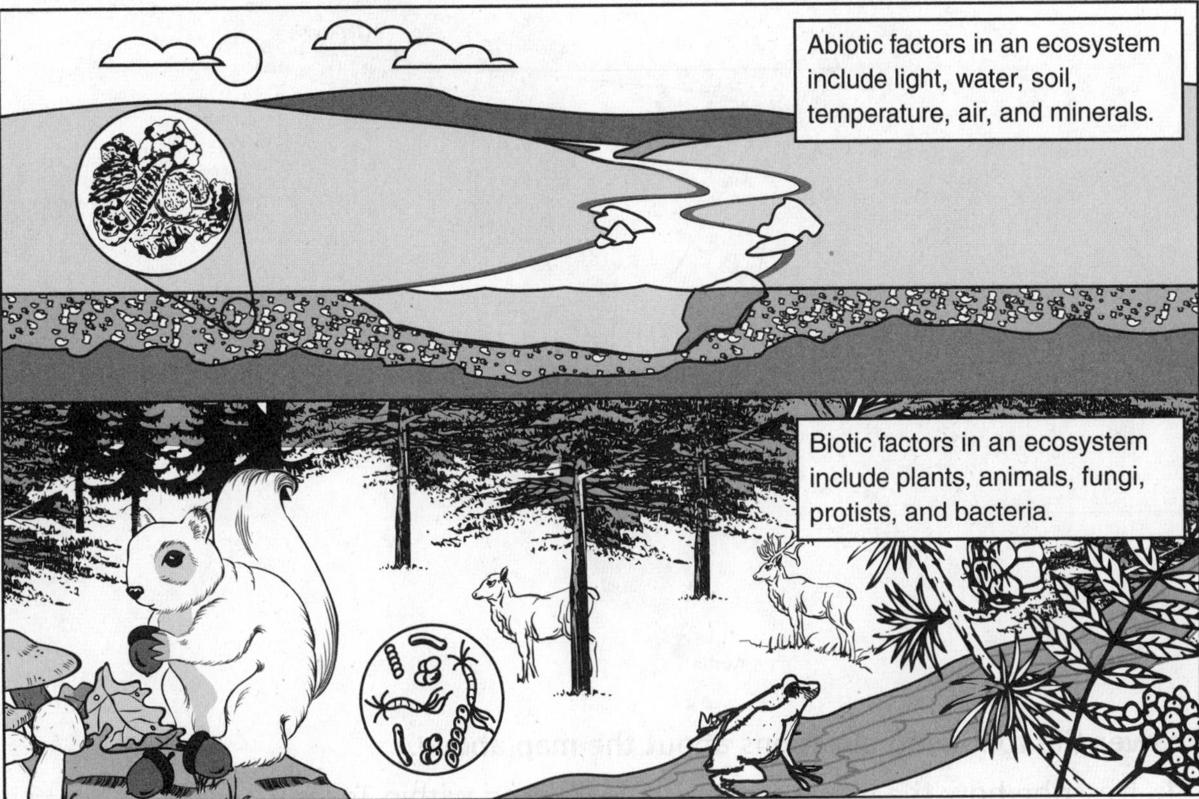

Abiotic factors in an ecosystem include light, water, soil, temperature, air, and minerals.
Biotic factors in an ecosystem include plants, animals, fungi, protists, and bacteria.

Answer these questions about the diagram above.

1. Name three biotic factors in the diagram.

____

2. Name three abiotic factors in the diagram.

____

3. What factors are in the biotic close-up?

____

4. What factors are in the abiotic close-up?

____

Name________________________ Date____________

Interpret Illustrations

Lesson 1

The Blackland Prairie

The Blackland Prairie is the largest remaining prairie in the United States. About 500 species of animals live on this prairie.

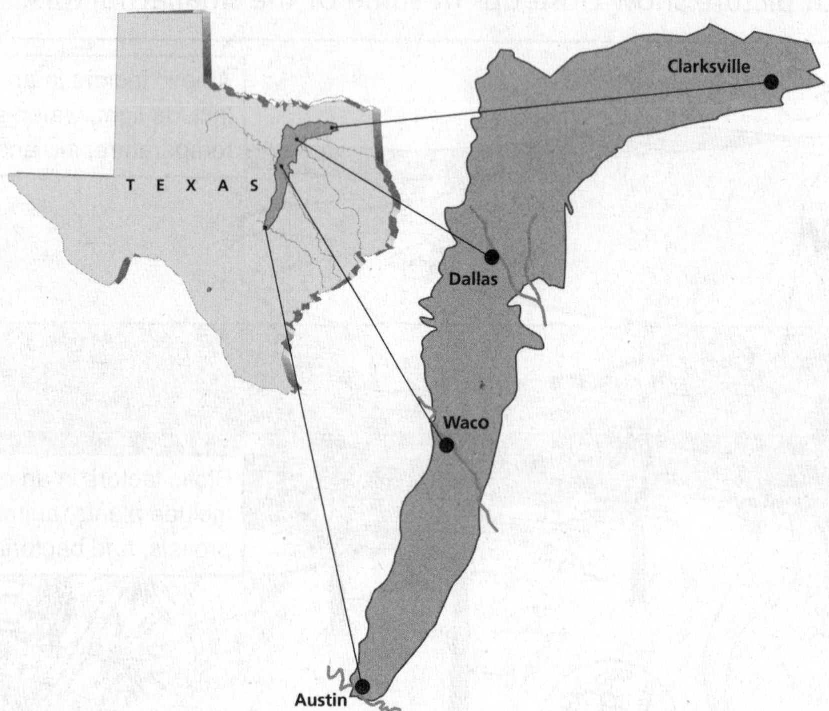

Answer the following questions about the map above.

1. Describe how the Blackland Prairie is situated within Texas.

____

____

2. Name three cities or towns within the Blackland Prairie.

____

____

3. What do most of the cities have in common in terms of the geographical features of their location?

____

____

Name ______________________ Date ____________

Lesson Vocabulary
Lesson 1

Interactions in an Ecosystem

Match the correct letter with the description.

Vocabulary
a. ecosystem
b. ecology
c. biotic factor
d. abiotic factor
e. population
f. community
g. habitat
h. niche

_____ 1. a living part of an ecosystem

_____ 2. all the populations living in an area

_____ 3. a nonliving part of an ecosystem

_____ 4. all the living and nonliving things in an area that interact with each other

_____ 5. the role of an organism in its community

_____ 6. all the organisms of one species that live in an area at the same time

_____ 7. the area in which an organism lives

_____ 8. the study of how living things and their environment interact

A new shopping center, which covers the soil habitat of worms, is built. Identify the following factors in an ecosystem as biotic or abiotic.

9. a parking lot ______________

10. worms ______________

11. soil ______________

12. new stores ______________

Answer the following questions.

13. What will happen to the worms when the parking lot is poured?

____

14. What is lost when the new stores are built? What is gained?

____

____

15. Birds eat worms. How does the new shopping center affect local birds?

____

Name______________________________ Date______________

Cloze Test
Lesson 1

Interactions in an Ecosystem

Vocabulary

ecosystem	minerals	dioxide	ecology
abiotic	water	energy	food

Fill in the blanks.

All the living and nonliving things in an area interacting with each other is a(n) ____________________. The study of how living things and their environment interact is called ____________________. The nonliving parts of an ecosystem are ____________________ factors. All living things require ____________________ to stay alive. Living things also need ____________________ such as iron and calcium. Algae and plants must have sunlight to make ____________________. They also need carbon ____________________. Animals require oxygen to produce the ____________________ their bodies need.

Use with textbook pages B4–B15

Name ______________________ Date ____________

Lesson Outline
Lesson 2

Interactions Among Living Things

Fill in the blanks.  Reading Skill: **Sequence of Events** - questions 2, 6, 20

What Is a Food Chain?

1. The path that energy takes from producers to consumers to decomposers is called a(n) ________________.
2. On a prairie, the food chains start with ______________, which produce food during ________________.

What Is a Food Web?

3. A food web shows the relationship between all ________________ in a community.
4. Organisms that use the Sun's energy to make their own food are called ________________.
5. Organisms that cannot make their own food are ________________.
6. Food chains end with ________________, which break down dead matter.

How Are Populations Connected?

7. A change in one ________________ affects all the other organisms in that food chain.
8. ______________________ help scientists predict how communities will be affected by change.

How Do Populations Adapt to Competition for Food?

9. Florida anoles were threatened when a new, bigger species of anole somehow arrived from ________________.
10. The Florida anole found a new ________________ high up in the treetops.

Name ______________________ Date __________

Lesson Outline

Lesson 2

What Is Symbiosis?

11. Symbiosis is the relationship between two kinds of ________________ over a period of time.

12. Sometimes both organisms ________________ from the relationship.

What Is Mutualism?

13. When both organisms benefit from the relationship, it is called ________________.

14. Yucca moths ________________ yucca flowers, which make seeds that sprout into ________________.

What Is Parasitism?

15. Parasitism is when one organism ________________ on or in another organism and may ________________ it.

16. A flea is a parasite that lives off the ________________ of a cat or a dog.

What Is Commensalism?

17. Commensalism is when one organism benefits from another without ________________ or ________________ it.

18. Orchids attach themselves to tree trunks to get ________________ without ________________ the trees.

How Does Energy Move in a Community?

19. Energy is ________________ as it passes from one organism to another in a food chain.

20. A(n) ________________ shows there is less food at the top than at the bottom.

How Do Food Webs Affect You?

21. The ocean turns red when a group of algae called ________________ bloom.

22. Poison from this algae can ________________ fish and make humans sick.

Name ______________________ Date ____________

What Is a Food Chain?

This illustration shows the members of a prairie food chain. You can see both producers and consumers. Identify where you think the food chain starts. Then follow the chain from one member to the next.

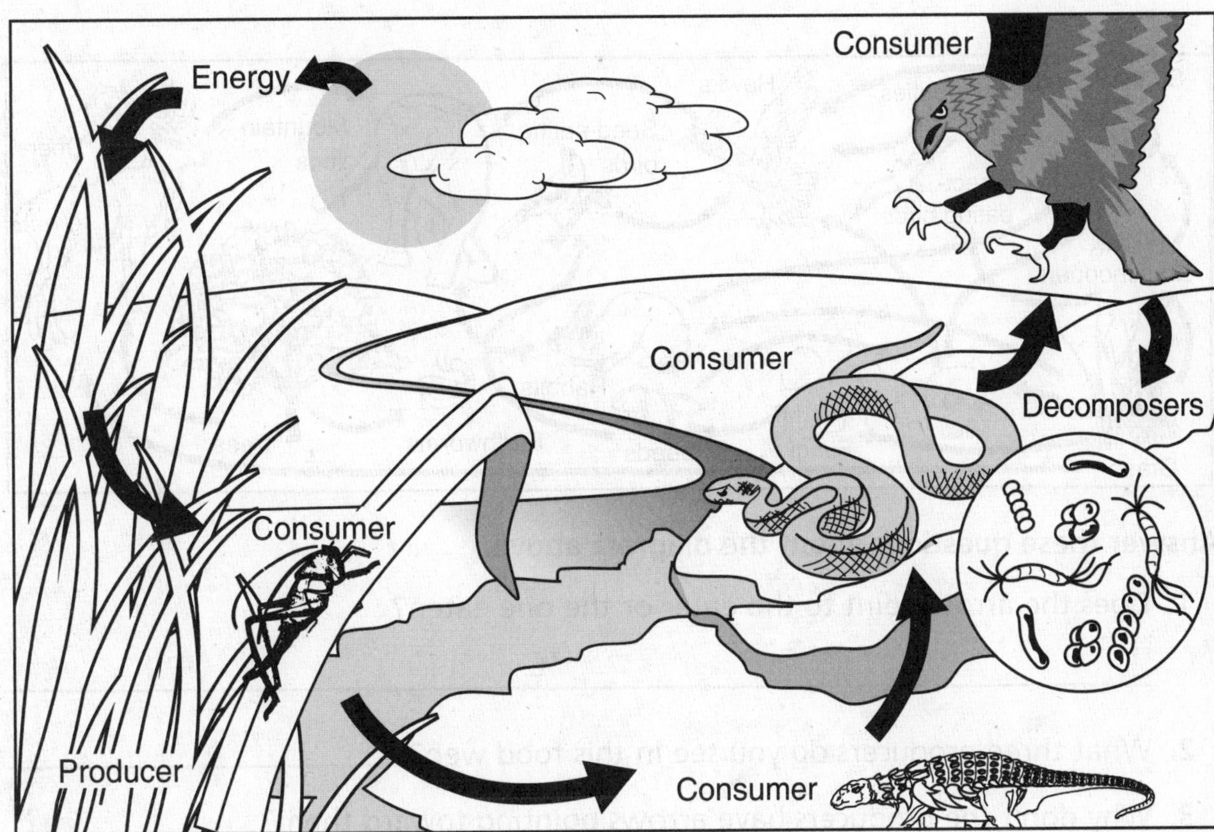

Answer these questions about the illustration above.

1. Where does the food chain start? ______________________
2. What producers are in the food chain? ______________________
3. What is the first consumer in the food chain? ______________________
4. What other consumers are in the food chain? ______________________
5. What is the direction of energy flow in this food chain?

____

6. What happens to the hawk? ______________________

Use with textbook pages B18–B19

Name ______________________ Date ____________

Interpret Illustrations

Lesson 2

What Is a Food Web?

This diagram shows a food web. Arrows point from an organism to the organisms that consume it. Notice that there is more than one arrow pointing to most of the consumers. Energy flows from producers to consumers. Trace the direction of several arrows to see how energy flows through the web.

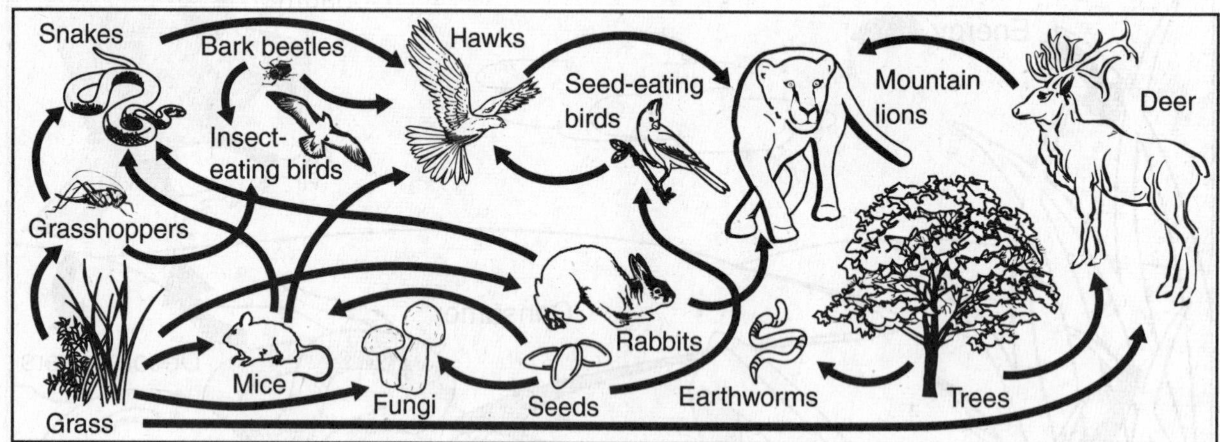

Answer these questions about the diagram above.

1. Does the arrow point to the eater or the one eaten?

   ____

2. What three producers do you see in this food web? ____________________

3. Why don't the producers have arrows pointing toward them?

   ____

   ____

4. Name three consumers you see.

   ____

5. What animals are not eaten by any other animal pictured in this food web?

   ____

6. What does the rabbit eat? What eats the rabbit?

   ____

Name ______________________________ Date ____________

Interactions Among Living Things

Match the correct letter with the description.

Vocabulary

- **a. food chain**
- **b. food web**
- **c. herbivore**
- **d. carnivore**
- **e. predator**
- **f. prey**
- **g. scavenger**
- **h. omnivore**
- **i. Sun**

_____ **1.** the overlapping food chains in a community

_____ **2.** a living thing that hunts other living things for food

_____ **3.** an animal that feeds on the remains of dead animals

_____ **4.** the original source of energy in an ecosystem

_____ **5.** the path of energy in food from one organism to another

_____ **6.** an animal that eats only other animals

_____ **7.** an animal that eats both plants and animals

_____ **8.** a living thing that is hunted for food

_____ **9.** an animal that eats only plants

Define the following and match the example(s) from the column of organisms.

10. producers ______________________________

____

11. consumers ______________________________

____

12. decomposers ______________________________

____

a. cows
b. humans
c. plants
d. bacteria

Answer the following questions.

13. Describe where plants, humans, and cows would be located on an energy pyramid.

top ____________ middle ____________ bottom ____________

14. Which level of the energy pyramid has the most organisms?

____

15. What would happen if suddenly all the cows disappeared?

____

____

Name ______________________ Date ____________

Cloze Test
Lesson 2

Interaction Among Living Things

Vocabulary

symbiosis | mutualism | parasitism
commensalism | relate | interactions

Fill in the blanks.

Organisms ________________ with one another in many different ways. The relationship known as ________________ is when two organisms interact with one another over a period of time. There are different ________________ that organisms have in this relationship. The yucca moth and yucca tree have a relationship based on ________________ because they both benefit from one another. In ________________ one organism feeds on another organism, which may be harmed in the relationship. Other organisms may have a relationship known as ________________ in which one organism benefits from the relationship without harming or helping the other.

Name ______________________ Date __________

Lesson Outline

Lesson 3

How Populations Survive

Fill in the blanks.  Reading Skill: **Summarize** - questions 2, 12

What Controls the Growth of Populations?

1. Anything that controls the growth or survival of a population is called a(n) ______________.
2. Limiting factors like temperature, sunlight, and water are ______________.
3. The number of predators an ecosystem can support is determined by the number of ______________.
4. A population's size can limit its ______________.

What Happens When Habitats Are Changed?

5. Bald eagles are found only in ______________.
6. As the human population settled all over North America, the bald eagle's natural habitat ______________, and their food supplies ______________.
7. In 1940 the ______________ was passed.
8. Laws were created to protect the bald eagle by banning ______________.
9. In 1995, the bald eagle's status was upgraded to ______________ species.

Name ______________________________ Date ______________

Lesson Outline

Lesson 3

How Do People Change the Environment?

10. In 2000 the ______________________ was 281,421,906 and the population per ______________________ was 79.6.

11. Air pollution is produced by ______________________ that are burned to power industries, transport systems, and homes.

12. To reduce the amount of garbage, people can apply the three Rs of conservation: ______________________, ______________________, and ______________________.

How Does Mining Change the Environment?

13. People use ______________________ in many ways.

14. Silver is used in ______________________ and ______________________.

15. Electricity flowing through ______________________ wires helps it run through appliances.

16. Another word for open-pit or strip mining is ______________________.

Name ______________________ Date ______________

Interpret Illustrations

Lesson 3

What Controls the Growth of Populations?

Limiting factors such as water, light, and food control the growth or survival of a population. The photograph below shows a group of walruses that is part of a larger population.

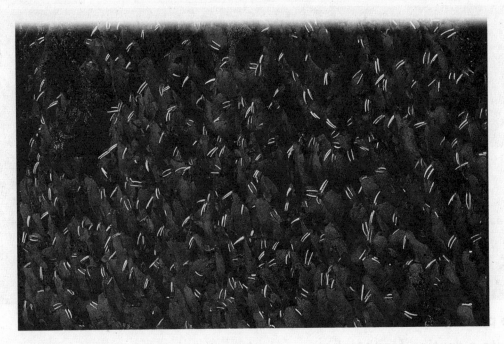

Use the photograph to answer the questions.

1. What are three non-living limiting factors that affect this group of walruses?

____

2. Describe two ways this population of walruses controls the populations of other organisms in the area.

____

____

____

____

3. How will this walrus population change when it exceeds the carrying capacity of the area?

____

Name ______________________ Date ____________

Lesson Outline

Lesson 3

How Does Mining Change the Environment?

The way people use land affects the environment. The photograph below shows how mining contributes to the degradation of the environment but without mining people would not be able to use resources to do everyday activities.

Use the photograph to answer the questions.

1. Write a caption for the photograph. A caption is a one or two sentence comment that summarizes or explains a picture or illustration.

____

____

2. What is surface mining?

____

____

3. Name three things that mining provides people with.

____

____

____

Use with textbook pages B40

Name ______________________ Date ____________

Lesson Vocabulary

Lesson 3

How Populations Survive

Match the correct letter with the description.

_____ 1. a species that has died out completely

_____ 2. a species in danger of dying out

_____ 3. the maximum population size that the resources in an area can support

_____ 4. anything that controls the growth or survival of a population

_____ 5. a species that may become endangered

Vocabulary

- a. limiting factor
- b. carrying capacity
- c. endangered species
- d. extinct
- e. threatened species

Identify the following as threatened, endangered, or extinct.

6. A dinosaur is a prehistoric organism. ____________

7. The bald eagle population has increased but still needs to be monitored. ____________

8. An animal's habitat has changed and its population has rapidly declined.

____

Answer the following questions.

9. Why was the American bald eagle placed on the endangered species list?

____

____

____

10. What is the difference between a species becoming threatened or endangered?

____

____

____

Name ______________________ Date __________

Cloze Test
Lesson 3

How Populations Survive

Vocabulary

population	**carrying capacity**	**compete**
limiting factor	**decrease**	
nonliving	**predators**	

Fill in the blanks.

A(n) ______________ is anything that controls the growth or survival of a(n) ______________. Some of these factors may be ______________ things, such as sunlight, wind, water, and temperature. The number of ______________ in an ecosystem also affects the number of prey. If there are more hawks than deer mice, there may be a(n) ______________ in the deer mouse population. Even a growing population faces problems. If the population becomes too crowded, the organisms will have to ______________ with one another for food, water, and shelter. The ______________ is the maximum population size that the resources in an area can support.

Use with textbook pages B32–B41

Name________________________ Date____________

Chapter Vocabulary

Chapter 5

Interactions of Living Things

Circle the letter of the best answer.

1. All the living and nonliving things in an area interacting with each other describes a(n)

a. community. **b.** ecosystem.

c. habitat. **d.** population.

2. The study of how living things and their environment interact is

a. biology. **b.** earth science.

c. ecology. **d.** geology.

3. A living part of a system is what kind of factor?

a. abiotic **b.** biotic

c. predator **d.** prey

4. All the organisms of one species that live in an area at the same time is a(n)

a. community. **b.** ecology.

c. ecosystem. **d.** population.

5. All the populations living in an area describes a(n)

a. community. **b.** ecosystem.

c. niche. **d.** population.

6. The area in which an organism lives is its

a. population. **b.** food web.

c. habitat. **d.** niche.

7. The role an organism has in its ecosystem is its

a. community. **b.** food chain.

c. habitat. **d.** niche.

Name ____________________ Date __________

8. The path of energy in food from one organism to another is a

a. food chain. b. food cycle.
c. food web. d. niche.

9. An animal that eats both plants and animals is a(n)

a. abiotic factor. b. carnivore.
c. herbivore. d. omnivore.

10. A living thing that is hunted for food is called

a. a carnivore. b. an omnivore.
c. a predator. d. prey.

11. An animal that feeds on the remains of dead animals is

a. an omnivore. b. a predator.
c. prey. d. a scavenger.

12. A relationship between two organisms that lasts over a period of time is

a. symbiosis. b. prey.
c. biotic. d. abiotic.

13. A relationship in which one kind of organism lives on or in another organism and may harm it is

a. biotic. b. mutualism.
c. parasitism. d. commensalism.

14. An animal that dies out completely is

a. endangered. b. threatened.
c. biotic. d. extinct.

Name_________________________ Date_____________

Chapter Summary

Chapter Summary

1. What are four vocabulary words you learned in the chapter? Write a definition for each.

2. What are two main ideas that you learned in this chapter?

Name______________________________ Date______________

Chapter Graphic Organizer
Chapter 6

Ecosystems

You can use a cycle diagram to summarize information. Each box gives important details. Follow the arrows to connect the ideas that summarize the carbon cycle. After one loop of the cycle has been completed, the cycle begins again.

The Carbon Cycle

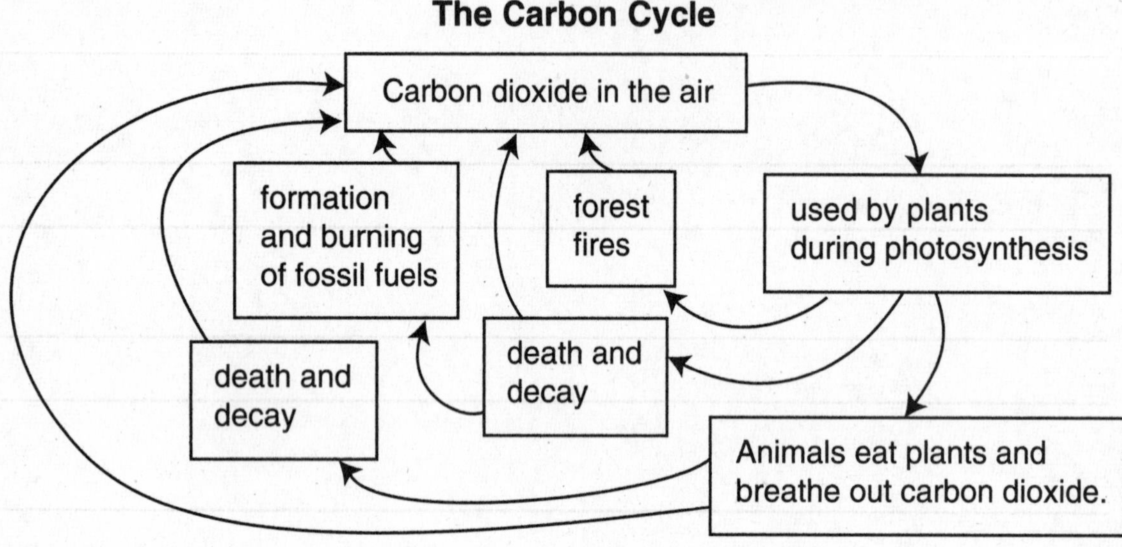

Complete the cycle diagram below for the water cycle. Two of the boxes have already been filled in.

The Water Cycle

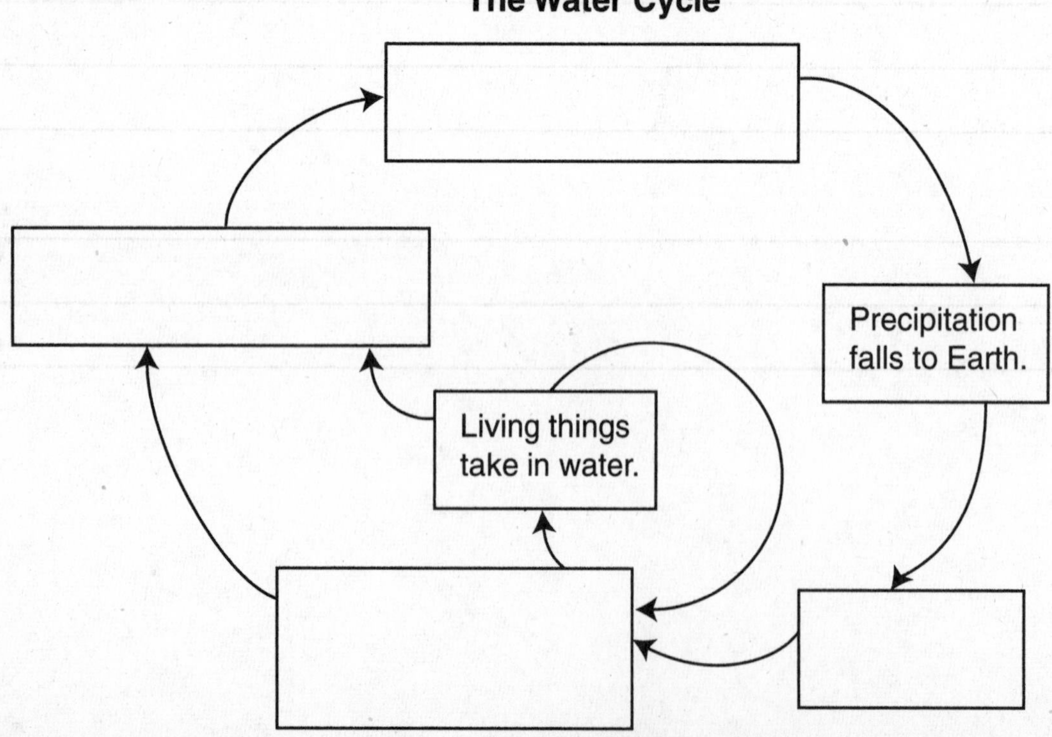

Name ______________________ Date __________

Chapter Reading Skill

Chapter 6

Summarize

A summary tells only the most important parts of a story. Usually, one or two sentences are enough to give a reader an idea of what the story will be about.

Read the following text based on information in your textbook. Then summarize the information in one or two sentences.

How Does Mining Change the Environment?

Metals play an important part in our modern society. However, we pay a price for them—and not only in money. Rocks that contain metal are often buried in the ground. We must change the ground to get them. If the rocks are near the surface, we carve away huge areas of land. This is called surface mining.

Surface-mined land is loaded with substances that are harmful to living things. Rainwater flows easily over this kind of land and carries pollutants into nearby streams, rivers, and lakes. The wind picks up dust, which pollutes the air. In both cases, living things are harmed.

____

____

____

____

____

____

Name________________________________ Date______________

Chapter Reading Skill
Chapter 6

Write a Summary

Read the following text based on information in your textbook. Then summarize it in one or two sentences.

Why Conserve?

The environment provides us with the resources needed to make products. Trees are made into paper for books, magazines, newspapers, and containers. The metals mined from the ground are used to make cars, ships, pots and pans, appliances, among many other things. Sand is made into glass. Plastic is made from chemicals in oil found deep underground.

Raw materials like oil and metals are nonrenewable resources. Once they are gone, they are gone forever. Trees and animals are renewable resources—they can be replaced. Even so, it takes time for trees and animals to grow. Fortunately, there are things we can do. Try practicing the three Rs of conservation: reduce, reuse, and recycle, as often as possible. For example, we can reduce the amount of gas and oil we use. Today, most cities and towns have recycling programs for paper, glass, and metal products. Learning how to conserve our resources is important for future generations.

My Summary:

____

____

____

____

____

____

Use with textbook pages B46–B93

Name________________________________ Date______________

Lesson Outline
Lesson 4

Cycles of Life

Fill in the blanks.  Reading Skill: **Sequence of Events** - question 13

What Is the Water Cycle?

1. Heat energy makes water ________________ and rise into the air.
2. When cooled enough, water vapor ________________ into tiny water droplets.
3. Droplets that become so large and heavy that they can no longer stay up in the air fall to Earth's surface as ________________.
4. Water stored in soil and rock is ________________.
5. Water flowing downhill across the surface of Earth is called ________________.
6. The process of naturally recycling water on Earth is called the ________________.
7. Plants give some of their water back to the atmosphere through their ________________.

What Is the Carbon Cycle?

8. The gas added to the air by decaying organisms, breathing animals, and burning fossil fuels is ________________.
9. Plants use carbon to make sugars, starches, and proteins during ________________________.
10. Animals use the carbon in sugars, starches, and proteins to make their own ________________________.

Name ______________________________ Date ____________

Lesson Outline
Lesson 4

How Is Nitrogen Recycled?

11. Animals get nitrogen by eating ____________________.
12. Plants get nitrogen by absorbing it from the ____________________.
13. The way nitrogen moves between the air, soil, plants, and animals is called the ____________________.
14. Decomposers break down the proteins of dead plants into the nitrogen-containing substance ____________________.
15. Soil bacteria change ammonia into ____________________.
16. Animal wastes contain ____________________ compounds.
17. Plants use nitrogen to make ____________________.
18. Nitrates are turned back into nitrogen gas by ____________________.

How Are Trees Recycled?

19. Dead trees provide ______________ that other trees need.
20. Organisms that recycle matter in dead organisms are called ______________.
21. Decomposers break down dead wood into carbon dioxide and ______________, which contains nitrogen.
22. Nitrogen is found in ______________, substances that add ______________ to the soil.

Why Recycle?

23. Sunlight is a(n) ____________________ resource.
24. The building blocks of products are ____________________.
25. A resource that can be replaced, like a tree, is ____________________.

Name ______________________ Date __________

Interpret Illustrations

Lesson 4

What Is the Carbon Cycle?

This illustration uses arrows to show how carbon cycles through the environment. Follow the arrows to see how carbon moves. Notice how the arrows form circles.

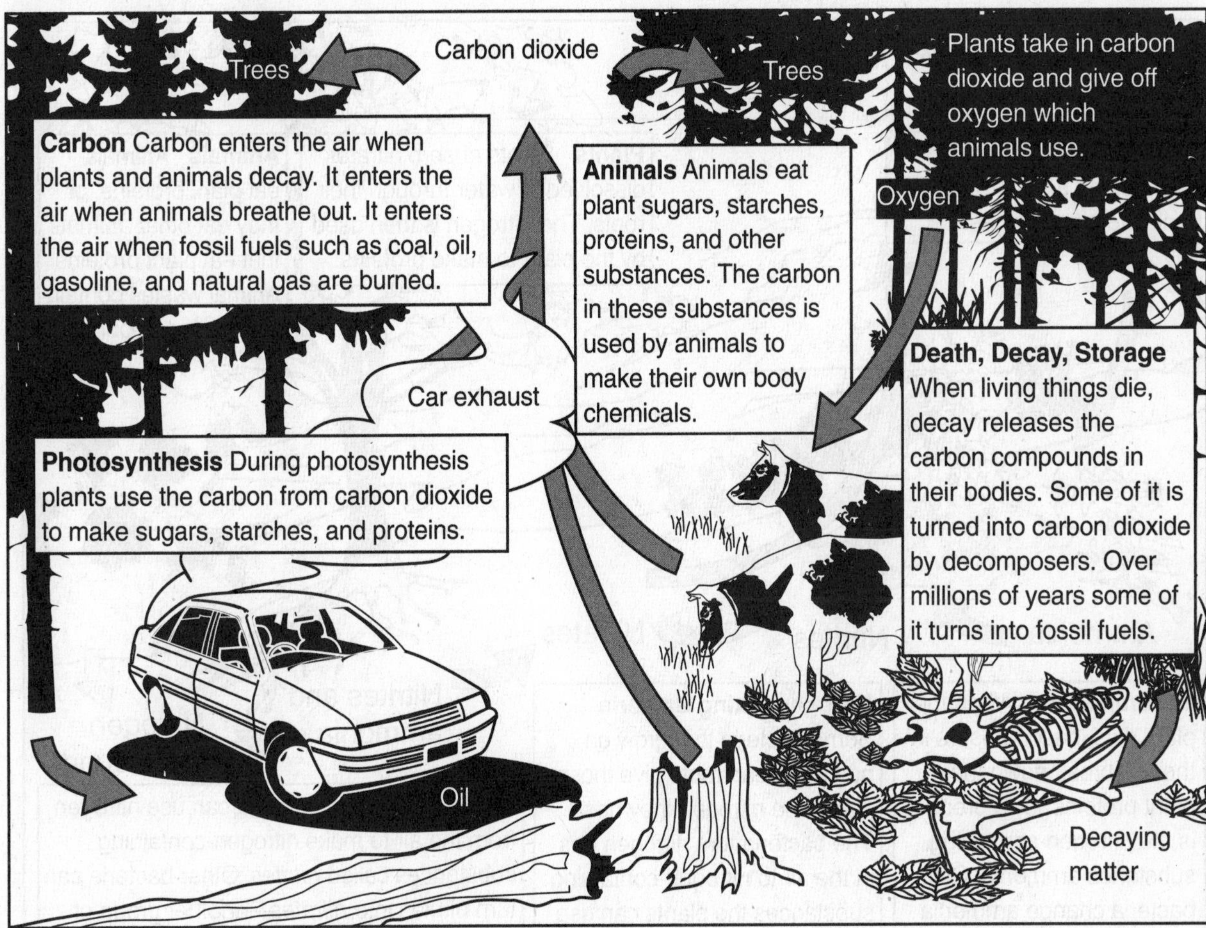

Trees
Carbon dioxide
Trees
Plants take in carbon dioxide and give off oxygen which animals use.
Carbon Carbon enters the air when plants and animals decay. It enters the air when animals breathe out. It enters the air when fossil fuels such as coal, oil, gasoline, and natural gas are burned.
Animals Animals eat plant sugars, starches, proteins, and other substances. The carbon in these substances is used by animals to make their own body chemicals.
Oxygen
Car exhaust
Death, Decay, Storage When living things die, decay releases the carbon compounds in their bodies. Some of it is turned into carbon dioxide by decomposers. Over millions of years some of it turns into fossil fuels.
Photosynthesis During photosynthesis plants use the carbon from carbon dioxide to make sugars, starches, and proteins.
Oil
Decaying matter

Answer these questions about the illustration above.

1. What sources of carbon dioxide are shown?

____

2. What takes in carbon dioxide? ______________________________

3. How do animals use carbon? ______________________________

4. Why is there an arrow from the plant to decaying matter?

____

5. Why does car exhaust release carbon dioxide? ______________________________

____

Name ____________________________ Date ______________

Interpret Illustrations

Lesson 4

How Is Nitrogen Recycled?

This diagram shows the nitrogen cycle. Begin at the cow and trace the arrows as you describe what is happening in your own words.

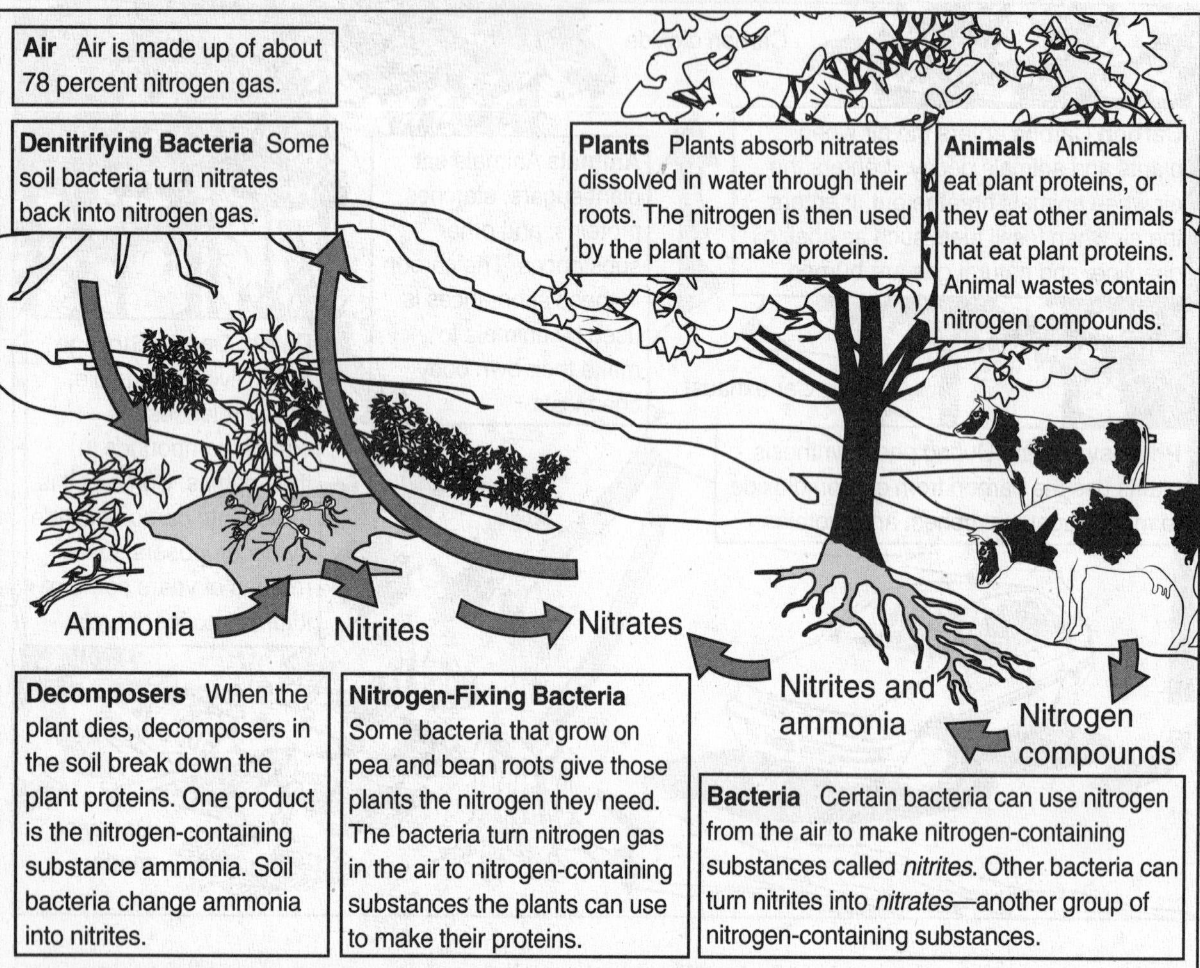

Air Air is made up of about 78 percent nitrogen gas.
Denitrifying Bacteria Some soil bacteria turn nitrates back into nitrogen gas.
Plants Plants absorb nitrates dissolved in water through their roots. The nitrogen is then used by the plant to make proteins.
Animals Animals eat plant proteins, or they eat other animals that eat plant proteins. Animal wastes contain nitrogen compounds.
Ammonia
Nitrites
Nitrates
Nitrites and ammonia
Nitrogen compounds
Decomposers When the plant dies, decomposers in the soil break down the plant proteins. One product is the nitrogen-containing substance ammonia. Soil bacteria change ammonia into nitrites.
Nitrogen-Fixing Bacteria Some bacteria that grow on pea and bean roots give those plants the nitrogen they need. The bacteria turn nitrogen gas in the air to nitrogen-containing substances the plants can use to make their proteins.
Bacteria Certain bacteria can use nitrogen from the air to make nitrogen-containing substances called *nitrites*. Other bacteria can turn nitrites into *nitrates*—another group of nitrogen-containing substances.

Answer these questions about the diagram above.

1. What nitrogen compounds can bacteria make? ____________________
2. Where do plants get nitrates from? ____________________

   ____

   ____
3. How does the cow rely on plants? ____________________

   ____
4. How does nitrogen leave an animal's body? ____________________
5. What organism turns nitrates back into nitrogen gas? ____________________

Name ______________________ Date __________

Cycles of Life

Vocabulary

a. evaporation
b. water vapor
c. condensation
d. precipitation
e. water cycle
f. carbon cycle
g. nitrogen cycle

Match the correct letter with the description.

_____ 1. any form of water that falls to Earth

_____ 2. the process in which a gas changes into a liquid

_____ 3. the continuous transfer of carbon between the atmosphere and living things

_____ 4. water in its gas state

_____ 5. the continuous movement of water between Earth's surface and the air

_____ 6. the transfer of nitrogen between the atmosphere and soil, plants, and animals

_____ 7. the process in which a liquid changes into a gas

Identify which cycle—water, nitrogen, or carbon—is linked to each term.

8. precipitation __________ 9. fertilizers __________

10. ammonia __________ 11. exhaust __________

12. bacteria __________ 13. condensation __________

14. photosynthesis __________ 15. oxygen __________

16. runoff __________ 17. evaporation __________

18. proteins __________ 19. fossil fuels __________

Describe one way animals are part of each of the three cycles.

20. water cycle ______________________________

____

21. carbon cycle ______________________________

____

22. nitrogen cycle ______________________________

____

Name ______________________ Date ____________

Cloze Test
Lesson 4

Cycles of Life

Vocabulary

oceans	hail	cycle	precipitation
Sun	runoff	clouds	groundwater

Fill in the blanks.

In nature, water continuously moves between Earth's surface and the air in a process called the water ____________. Most of Earth's water is contained in ____________. Energy from the ____________ evaporates water, making it rise. Some water droplets and ice crystals form ____________. When these grow large and heavy, the water falls to Earth's surface as ____________. This can take the form of snow, rain, ____________, or sleet. Water can be stored as ____________ or it can flow as ____________.

Name ______________________ Date __________

Lesson Outline
Lesson 5

Biomes

Fill in the blanks.  Reading Skill: **Summarize** - questions 10, 22

What Is a Biome?

1. The six major kinds of large ecosystems are called ______________.
2. Rich topsoil over clay is found in the ______________.
3. Acidic soil with a surface of decayed pine needles describes the ______________.
4. A biome with nutrient-poor soil near the equator is the ______________ ______________.
5. You would find a soil poor in animal and plant decay products, but rich in minerals in a(n) ______________.
6. Nutrient-poor soil and permafrost are found in the ______________.
7. Rich topsoil and few trees characterize a(n) ______________.

What Are Grasslands?

8. Prairies and savannas are two types of ______________.
9. Grasslands in the United States are ______________, meaning mild.
10. Grasslands are called the ______________ of the world because they produce so much wheat, corn, and oats.

What Is the Taiga Like?

11. Lakes and ponds in the taiga were formed by moving ______________.
12. Taigas are mostly ______________ forests.

What Is the Tundra?

13. The biome located between the taiga and the polar ice sheets is the ______________.
14. Permanently frozen soil, or ______________, keeps water from flowing downward.

Name ______________________________ Date ______________

Lesson Outline

Lesson 5

What Is the Desert Biome Like?

15. A sandy biome with little precipitation and plant life is a(n) ______________.
16. Every continent has a desert, but the ______________________ in Africa is the largest.

What Is a Deciduous Forest?

17. Many trees lose their leaves each year in a(n) ______________________.
18. The word "deciduous" means ______________________ to the ground.

What Are Tropical Rain Forests?

19. Tropical rain forests are near Earth's ______________________.
20. Little sunlight reaches the ground in a tropical rain forest because of the thick ______________________.
21. Most of the life in a tropical rain forest is high up in the ______________.

What Are Water Ecosystems Like?

22. The main difference between Earth's water ecosystems is ______________.
23. Organisms that float on the water are called ______________________.
24. Organisms that swim through the water are ______________________.
25. Organisms that dwell on the bottom are ______________________.
26. Crabs, mussels, and barnacles live in the ______________________ zone.
27. Photosynthetic organisms cannot live in the ______________________ region of the ocean because there is no sunlight.

Can Humans Change Water Ecosystems?

28. Many species of whales are threatened with ______________________.
29. Whales were used for food, oil, and ______________________.
30. The United States banned citizens from buying products made from whales in ______________________.

Name ______________________ Date __________

Interpret Illustrations

Lesson 5

What Is a Biome?

Maps like this one show one aspect of Earth. In this case, biomes are shown in colors or textures. Make sure you look at how each biome is indicated. Then read the map to see where various biomes can be found.

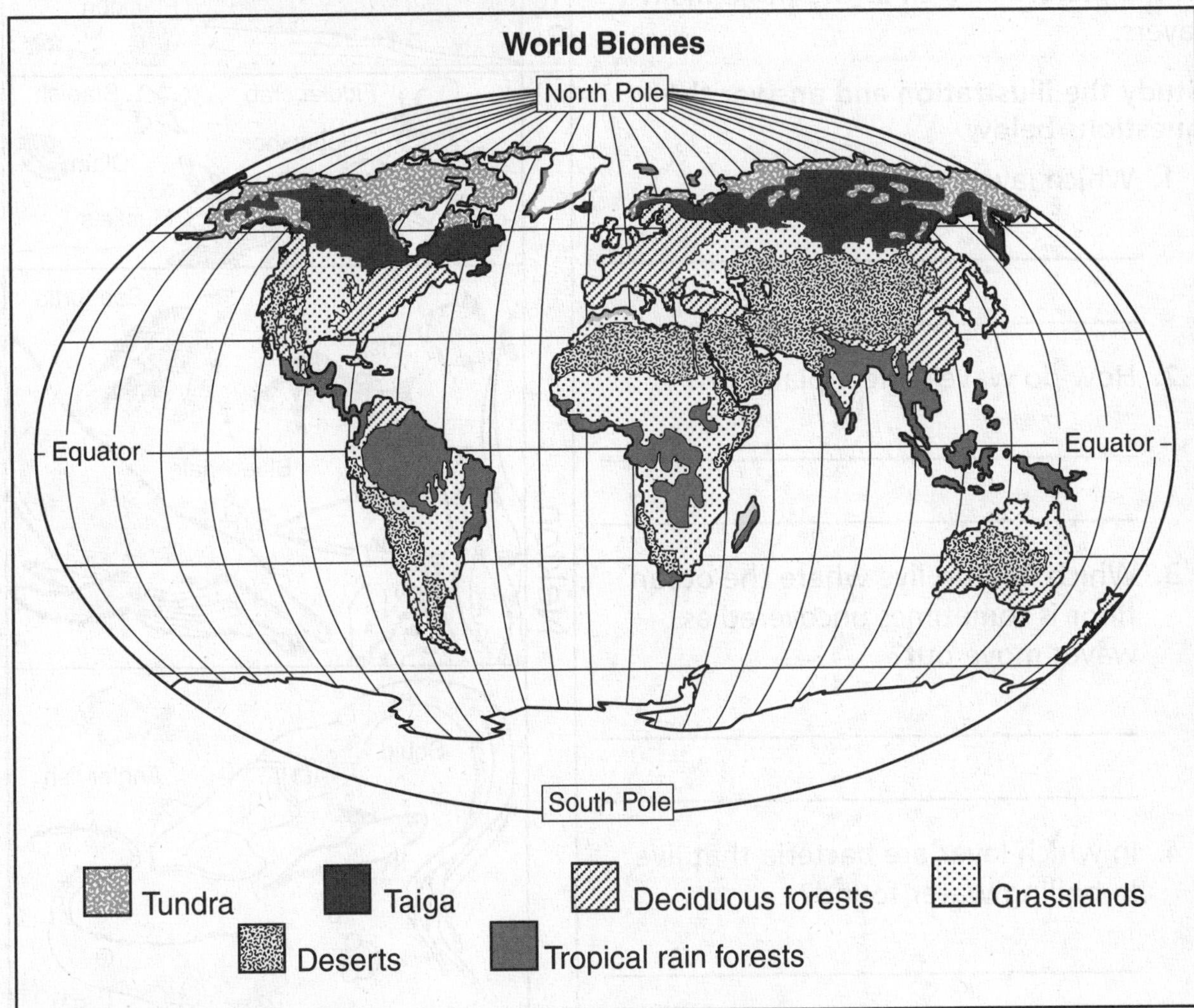

Study the map of biomes and answer the questions below.

1. What biome is found mostly near the equator? ______________________
2. What biome stretches across northern Africa? ______________________
3. What biome is found just below the North Pole? ______________________
4. What are the three main biomes found in the United States?

____

5. What biome do you live in? ______________________

Name ______________________ Date ____________

Interpret Illustrations

Lesson 5

What are Water Ecosystems Like?

This illustration shows the kind of life at different levels of the ocean. Make sure you can identify each of the organisms in the picture as well as the three main layers.

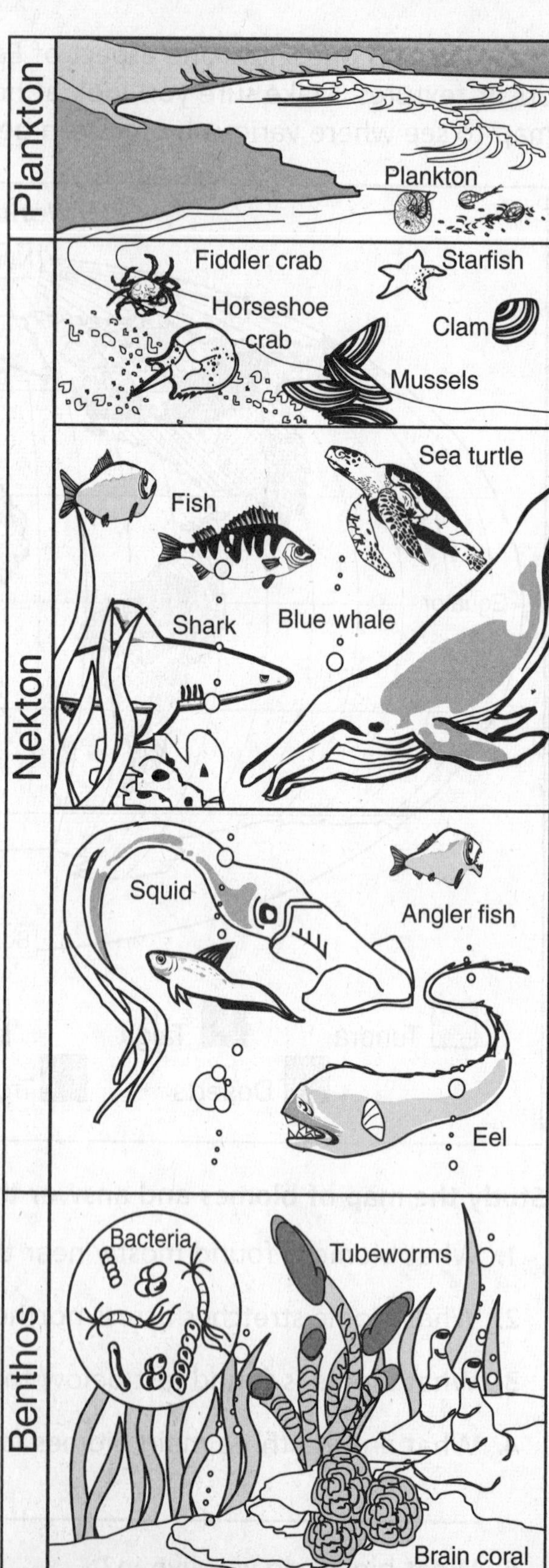

Study the illustration and answer the questions below.

1. Which layer has waves?

____

____

2. How do waves affect plankton?

____

____

3. Which nekton live where the ocean floor is sometimes uncovered as waves move out?

____

____

4. In which layer are bacteria that live in boiling water found?

____

____

5. Which organisms get more sunlight—plankton or the bacteria that live in boiling water?

____

Use with textbook page B73

Name ______________________________ Date ______________

Lesson Vocabulary

Lesson 5

Biomes

Match the correct letter with the description.

_____ **1.** a forest biome with many kinds of trees that lose their leaves each autumn

_____ **2.** one of Earth's large ecosystems, with its climate, soil, plants, and animals

_____ **3.** a hot, humid biome near the equator, with much rainfall and a wide variety of life

_____ **4.** a cold, treeless biome of the far north, marked by spongy topsoil

_____ **5.** most lakes, streams, rivers, and ponds

_____ **6.** a sandy or rocky biome, with little precipitation and little plant life

_____ **7.** a cool, forest biome of conifers in the upper Northern Hemisphere

_____ **8.** have tides, an upper region with many fish and whales, and a dark bottom

Vocabulary

a. biome
b. taiga
c. tundra
d. desert
e. deciduous forest
f. tropical rain forest
g. freshwater ecosystems
h. saltwater ecosystems

Identify the biome described by each of the following.

9. few plants live on its floor because little sunlight penetrates the thick canopy ______________________

10. plants grow deep roots to find scarce water ______________________

11. small plants have shallow roots and short growing seasons because of the permafrost ______________________

12. rainfall is irregular and much of this biome has been turned into farmlands ______________________

13. mostly conifer trees in a cool climate ______________________

14. leaves turn color and fall during autumn ______________________

15. has boiling vents on the dark floor ______________________

16. where frogs, turtles, and brook trout live ______________________

Name______________________________ Date______________

Cloze Test

Lesson 5

Biomes

Vocabulary

minerals	leaves	tundra	taiga
permafrost	grasslands	plants	biomes

Fill in the blanks.

Six major ecosystems, or ____________________, are located around Earth. The desert has soil that is rich in ____________________. Prairie dogs and snakes live in American ____________________. The cold biome in the far north is the ____________________. Here, a layer of permanently frozen soil called ____________________ keeps water from flowing downward. The greatest diversity of ____________________ can be found in tropical rain forests. Mostly conifers grow in the ____________________. Deciduous forests have many different trees that lose their ____________________ each autumn.

Name________________________ Date____________

Lesson Outline

Lesson 6

How Ecosystems Change

Fill in the blanks.  Reading Skill: **Summarize** - questions 4, 10

How Do Ecosystems Change?

1. ____________________ has a way of changing an ecosystem or producing a new one.
2. In Cambodia, abandoned cities have turned back into ____________________.
3. Over many years, an abandoned farm can turn into a(n) ____________________.

How Do Communities Change?

4. The gradual replacement of one community by another is ____________________.
5. Ecological succession in a place where a community already exists is called ____________________.
6. Where communities have been wiped out, ____________________ can begin.
7. A year after Mount Saint Helens erupted, you could find the rose-purple ____________________ of fireweed in the rubble.
8. The first species to be living in an otherwise lifeless area is a(n) ____________________.
9. A new community is called a(n) ____________________.

What Happens to Pioneer Communities?

10. Over many years, ____________________, ____________________, and ____________________ can break down rock into soil.
11. The final stage of succession is the ____________________.
12. After a major event, the processes of ____________________ begin all over again.

Name________________________________ Date______________

Lesson Outline

Lesson 6

What's Living on Surtsey?

13. Surtsey is off the coast of ____________________.
14. Surtsey was formed by a(n) ____________________ in 1963.
15. Between 1963 and 1996, at least ____________________ types of plants were growing on Surtsey.

How Do Populations Survive Earth's Changes?

16. Some organisms survive major changes on Earth, while others become ____________________.
17. Scientific evidence suggests that a long time ago a(n) ____________________ struck Earth, which caused many plants to die.

What Do Fossils Tell Us About Changes in the Environment?

18. About six million years ago, fish and other sea creatures disappeared from the ____________________.
19. Fossils from a slightly later period reveal that ____________________ from Africa arrived in Europe.
20. Fossils of fish from ____________________ years ago turned up in the Mediterranean area.
21. Scientists think that Africa and Europe bumped each other at the ____________________.
22. This collision created a natural ____________________ between the Atlantic Ocean and the Mediterranean Sea.
23. The sea became a(n) ____________________ and fish and other marine life died out.
24. About five million years ago, the dam crumbled and water, carrying ____________________, made the Mediterranean a sea again.

Name______________________________ Date____________

Interpret Illustrations

Lesson 6

How Do Ecosystems Change?

These five pictures show the same place at five different times. By comparing pictures you can see how the environment has changed over a period of 100 years. Look at each picture and identify the different plants and animals.

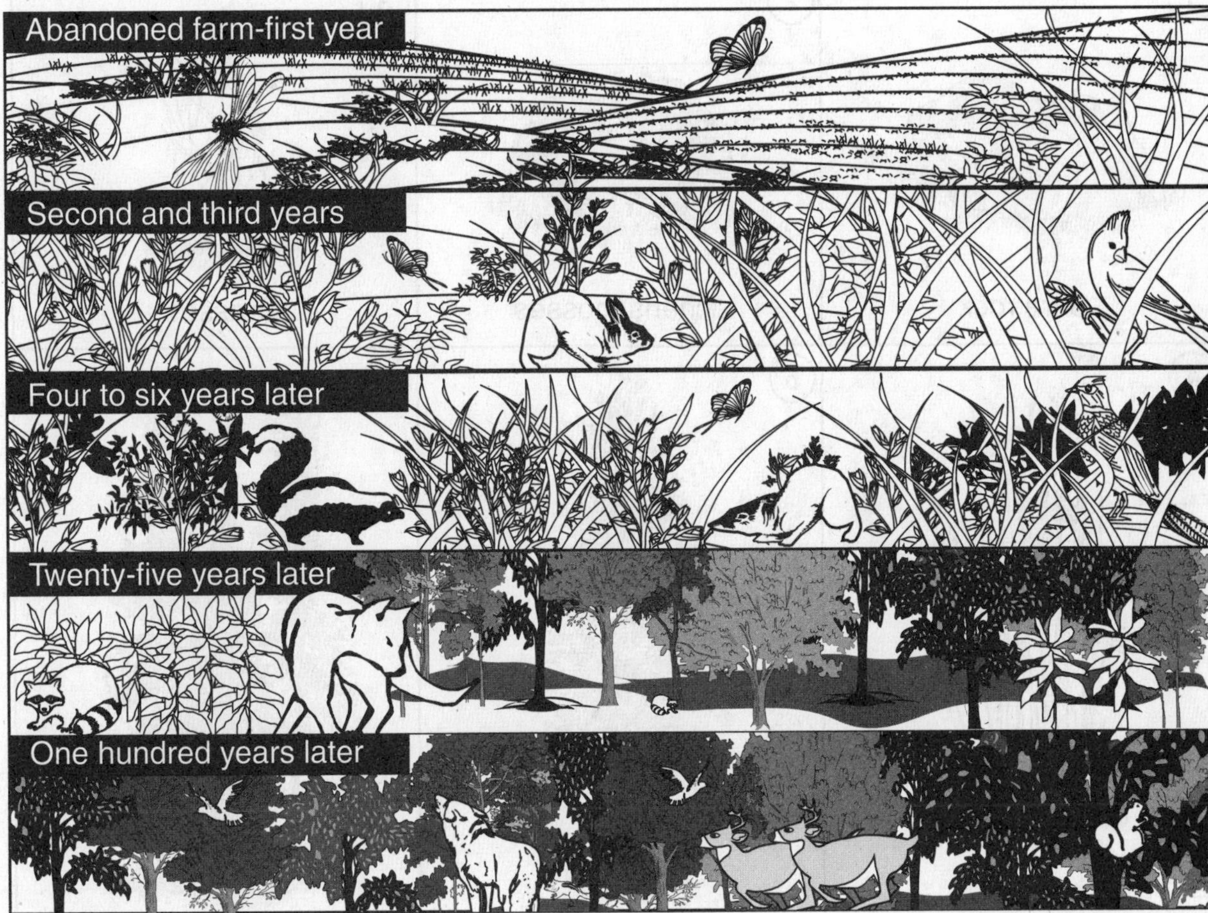

Use the pictures above to answer these questions.

1. What kinds of plants do you see in the first picture? ____________

2. What happens to these plants? ____________

3. How do the animals change through the years? ____________

____

4. How do the plants change through the years?

____

5. How many years pass before birds come to the area? ____________

Name_____________________________ Date______________

Interpret Illustrations

Lesson 6

Stages of Succession

These pictures focus on how a small part of an ecosystem changes over time. The pictures show how one thing gradually replaces another.

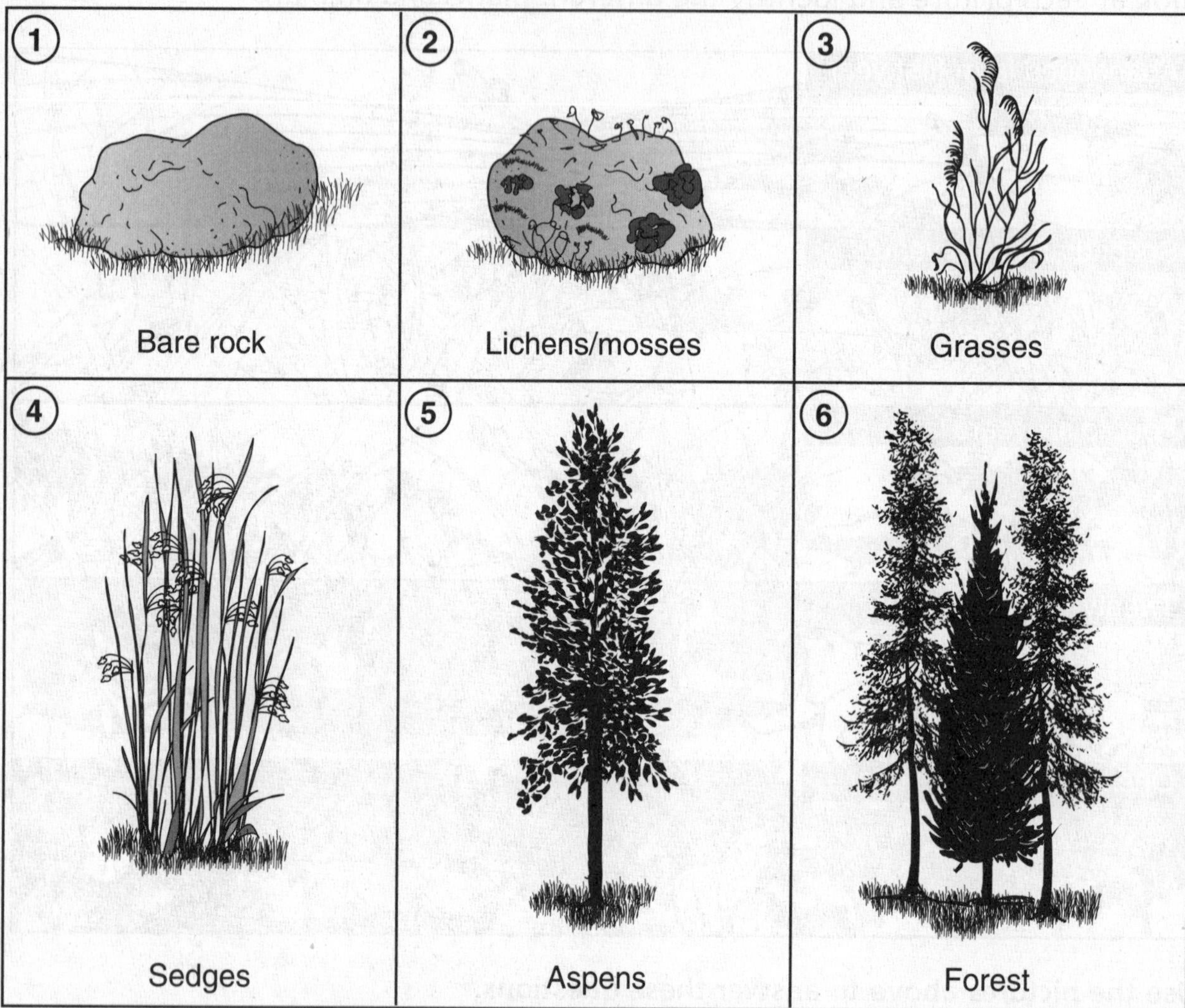

Use the pictures above to answer these questions.

1. What is the item in the first picture? ______________________

2. What is the difference between the rock in the first and second pictures?

____

3. What happens to the rock? ______________________

4. What do the aspens replace? ______________________

5. What is shown in the final picture? ______________________

Name ______________________ Date __________

Lesson Vocabulary
Lesson 6

How Ecosystems Change

Match the correct letter with the description.

_____ **1.** the first species living in an area

_____ **2.** the gradual replacement of one community by another

_____ **3.** ecological succession in a place where a community already exists

_____ **4.** the first community living in an area

_____ **5.** ecological succession that happens where there are few, if any, living things

_____ **6.** the final stage of succession in an area, unless a major change happens

Vocabulary

a. ecological succession
b. pioneer species
c. pioneer community
d. climax community
e. primary succession
f. secondary succession

Complete the table by listing one early change and one later change that took place following each event.

What Happened	Early Change	Later Change
farm abandoned	**7.**	**8.**
volcanic eruption	**9.**	**10.**
new building	**11.**	**12.**
continents bump	**13.**	**14.**

Answer the following questions.

15. How can changes in one ecosystem affect another ecosystem?

____

16. Name three things you can do to help ecosystems.

____

Name____________________ Date__________

Cloze Test
Lesson 6

How Ecosystems Change

Vocabulary

forest	pioneer	secondary succession	eruption
primary succession	succession	fire	pioneer

Fill in the blanks.

The gradual replacement of one community by another is called ecological ____________. When a community begins where another community already exists, ____________ occurs. For example, a(n) ____________ can gradually replace a farm. When a community begins where there are few, if any, living things, ____________ occurs. Such places include land swept clean by a volcanic ____________ or forest ____________. The first species to be living in an otherwise lifeless area is a(n) ____________ species. A new community is also known as a(n) ____________ community.

Name ______________________ Date __________

Chapter Vocabulary

Chapter 6

Ecosystems

Circle the letter of the best answer.

1. A process in which a liquid changes into a gas is
   a. condensation. b. evaporation.
   c. precipitation. d. succession.

2. Water droplets that fall to Earth from clouds is
   a. evaporation. b. condensation.
   c. precipitation. d. symbiosis.

3. Clouds are formed from the process of
   a. condensation. b. mutualism.
   c. precipitation. d. evaporation.

4. Evaporation, condensation, and precipitation are a part of the
   a. carbon cycle. b. nitrogen cycle.
   c. water cycle. d. oxygen cycle.

5. The gas that moves in a cycle between the air, soil, plants, and animals is
   a. oxygen. b. nitrogen.
   c. neon. d. hydrogen.

6. One of Earth's large ecosystems, with its climate, soil, plants, and animals, is a
   a. biome. b. climax community.
   c. pioneer community. d. taiga.

7. A cool, forest biome of conifers in the upper Northern Hemisphere is a
   a. deciduous forest. b. taiga.
   c. tropical rain forest. d. tundra.

8. A sandy or rocky biome, with little precipitation and little plant life, is a
   a. desert.
   b. pioneer community.
   c. taiga.
   d. tundra.

9. A forest biome with many kinds of trees that lose their leaves each autumn is a
   a. deciduous forest.
   b. taiga.
   c. tropical rain forest.
   d. tundra.

10. A hot, humid biome near the equator, with much rainfall and a wide variety of life, is a
   a. deciduous forest.
   b. taiga.
   c. tropical rain forest.
   d. tundra.

11. The gradual replacement of one community by another is called
   a. commensalism.
   b. ecological succession.
   c. pioneer succession.
   d. symbiosis.

12. The first species living in an area is the
   a. climax species.
   b. ecological succession.
   c. pioneer species.
   d. taiga.

13. The first community living in an area is the
   a. climax community.
   b. ecological community.
   c. pioneer community.
   d. premier community.

14. The final stage of succession in an area, unless a major change happens, is the
   a. climax community.
   b. ecological community.
   c. pioneer community.
   d. secondary community.

Name ______________________ Date ____________

Unit Vocabulary

Unit B

Name That Word

Use the words in the box to fill in the blanks.

Word Box

carnivore	population	biotic factor	herbivore
carbon cycle	niche	biome	food chain
ecological succession	prey	habitat	scavenger

1. These words identify a living part of an ecosystem.

   ____

2. These words identify the movement of food energy from one organism to another. ______________________

3. This word identifies an organism's role in an ecosystem.

   ____

4. This word identifies a plant eater. ______________________

5. This word identifies all the members of a species in an area.

   ____

6. These words identify the transfer of carbon between the atmosphere and living things. ______________________

7. This word identifies a meat eater. ______________________

8. This identifies the gradual replacement of one community by another.

   ____

9. This word identifies a meat eater that eats the remains of dead animals.

   ____

10. This word identifies a living thing hunted as food. ______________________

11. This word identifies where a population lives. ______________________

12. This word identifies a large ecosystem with its own kind of climate, soil, plants, and animals. ______________________

Name ______________________ Date ____________

Unit Vocabulary

Unit B

Correct-a-Word

One word in each sentence below is wrong. Cross it out, and write the correct word above it. Use the Word Box to check your spelling.

Word Box				
deciduous	precipitation	ecology	decomposer	nitrogen
ecosystem	scavengers	pioneer	habitat	niche

1. Noxrogen is an element that all plants need for growth.

2. A primitive species is the first species to be living in an otherwise lifeless area.

3. Scallywags are meat eaters that eat the remains of dead animals.

4. An echovalley includes all the living and nonliving things in an environment.

5. The place where a population lives is a haberole.

6. Its nick is an organism's role in an ecosystem.

7. A producer is fungus or bacterium that breaks down dead plants and animals.

8. Deciding trees lose leaves in fall.

9. Precession is the rain, sleet, snow, or hail that falls to Earth.

10. Economy is the study of the interaction of living things and their environment.

Name________________________ Date____________

Unit Vocabulary

Unit B

Find-a-Word

Look across, down, and diagonally to find these hidden words:

ABIOTIC, ADAPT, AMMONIA, ANIMALS, AREA, BENTHOS, BIOME, CARBON, CARNIVORE, CHAIN, COMMUNITY, CONSUMER, DECIDUOUS, DESERT, EAT, ECOLOGY, ECOSYSTEM, FOOD, GAS, GRASSLAND, GROWTH, HERBIVORE, NEKTON, NICHE, NITROGEN, OMNIVORE, PARASITE, PLANKTON, PLANTS, POPULATION, PREY, PRODUCER, ROLE, SCAVENGER, SOIL, SPECIES, SURVIVE, SYMBIOSIS, TAIGA, TIDE, TUNDRA, WEB.

A	D	A	P	T	Z	S	Y	M	B	I	O	S	I	S	A
E	C	O	L	O	G	Y	Q	A	B	I	O	T	I	C	N
Q	C	C	Z	J	D	E	C	I	D	U	O	U	S	A	I
P	R	O	D	U	C	E	R	S	O	I	L	M	Q	V	M
O	S	N	S	G	R	A	S	S	L	A	N	D	E	E	A
P	U	S	I	Y	E	Q	B	E	N	T	H	O	S	N	L
U	R	U	P	C	S	A	Z	D	E	S	E	R	T	G	S
L	V	M	A	A	H	T	T	U	N	D	R	A	W	E	B
A	I	E	R	R	F	E	E	C	A	R	B	O	N	R	A
T	V	R	A	E	C	O	M	M	U	N	I	T	Y	G	M
I	E	Z	S	A	P	R	O	L	E	E	V	A	C	R	M
O	M	N	I	V	O	R	E	D	T	K	O	I	H	O	O
N	N	I	T	R	O	G	E	N	I	T	R	G	A	W	N
Z	S	P	E	C	I	E	S	Y	D	O	E	A	I	T	I
C	A	R	N	I	V	O	R	E	E	N	J	S	N	H	A
P	L	A	N	K	T	O	N	J	X	P	L	A	N	T	S

Name__________________________ Date____________

Chapter Graphic Organizer
Chapter 7

Rocks and Minerals

You can use a flowchart to show a sequence of events, such as the erosion of rock by ice. The first event is written in a box. The second event is written in a second box, and so on until all the events are listed. The boxes are then connected with arrows to show the order in which the events occur. Look at the example below showing how ice erodes rock.

Erosion of Rock by Ice

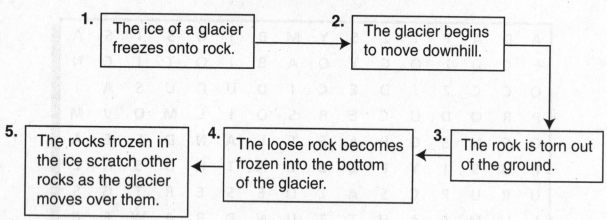

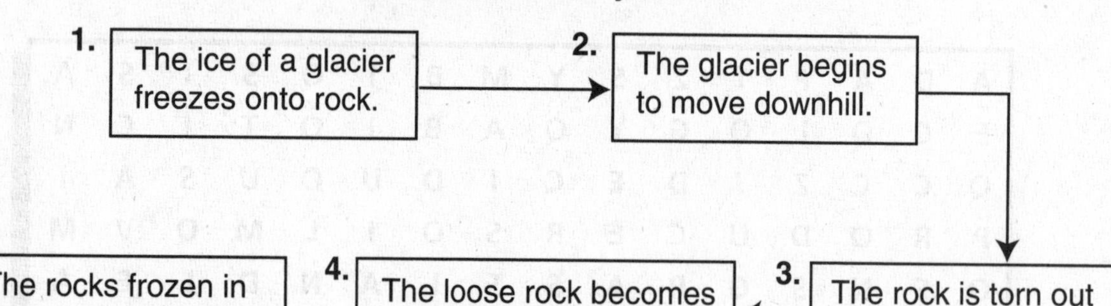

1. The ice of a glacier freezes onto rock.
2. The glacier begins to move downhill.
3. The rock is torn out of the ground.
4. The loose rock becomes frozen into the bottom of the glacier.
5. The rocks frozen in the ice scratch other rocks as the glacier moves over them.

A flowchart can help you see how a fossil is formed in sedimentary rock. **Complete the following flowchart to show the sequence of events.** Two of the boxes have already been filled for you.

Steps in the Formation of a Fossil

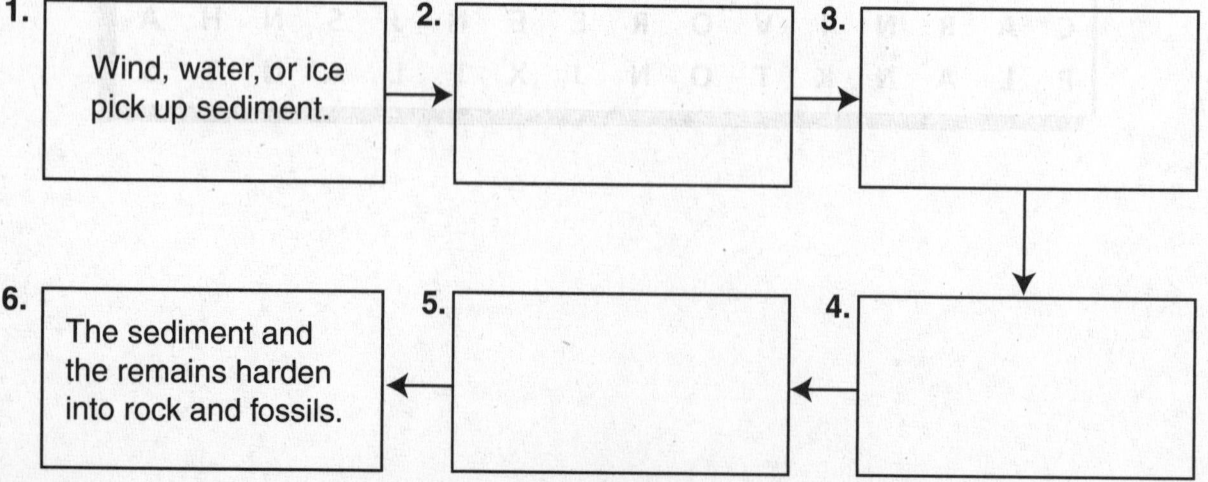

1. Wind, water, or ice pick up sediment.

6. The sediment and the remains harden into rock and fossils.

Name ______________________ Date ____________

Sequence of Events

Life is a series of sequences, such as eating dinner, getting dressed, taking a trip, playing baseball, and going to bed. Sometimes you have to decide what you can do based on sequence. Look at the camp schedule below. It lists activities and the times they are available. **Use information from the schedule to answer the questions and make your own sequence of events.**

Camp Lotsafun Schedule

8:00 A.M. All campers arrive by bus.
8:30 A.M.-12:30 P.M. Activities
12:30-1:00 P.M. Lunch is available.
1:00-1:30 P.M. Lunch is available.
1:30-4:00 P.M. Activities
4:15 P.M. Buses leave.

Acting

9:00-10:00 A.M.
11:00 A.M.-noon
12:30-1:30 P.M.
1:30-2:30 P.M.
3:00-4:00 P.M.

Boating

9:00-10:30 A.M.
11:00 A.M.-12:30 P.M.
2:00 A.M.-3:30 P.M.

Crafts

8:30-9:30 A.M.
10:30-11:30 A.M.
1:30-2:30 P.M.
2:30-3:30 P.M.

Gymnastics

8:30-9:00 A.M.
10:00-10:30 A.M.
noon-12:30 P.M.
1:30-2:00 P.M.
3:00-3:30 P.M.

Horseback Riding

9:00-10:00 A.M.
10:00-11:00 A.M.
11:30-12:30 P.M.
1:00-2:00 P.M.
2:00-3:00 P.M.
3:00-4:00 P.M.

Soccer

9:00-11:30 A.M.
11:00 A.M.-1:00 P.M.
2:00-4:00 P.M.

1. What time do you get to camp? ____________
2. What time do you go home? ____________
3. When is lunch available? ____________
4. Make a schedule that allows you to take acting, gymnastics, and boating *or* horseback riding, crafts, and soccer and still have lunch!

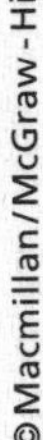

Name ______________________ Date ____________

Chapter Reading Skill
Chapter 7

One Step at a Time

We do everything in sequence, one step at a time. First you use dishes, then you wash them, dry them, and put them away. When you get a glass of milk or spend a day in school, you follow a step-by-step sequence.

Number the entries on each list below to show the sequence of events. Add entries if you want flavored milk or do other things at school.

Get a Glass of Milk	My Day at School
_____ Drink the milk. Yum!	_____ Go home from school.
_____ Dry the glass.	_____ Have gym class.
_____ Pour the milk.	_____ Have math class.
_____ Put the glass away.	_____ Arrive at school.
_____ Get milk from the refrigerator.	_____ Have science class.
_____ Wash out the glass.	_____ Have lunch.
_____ Put milk back in the refrigerator.	_____ Have social studies class.
_____ Get a drinking glass.	_____ Hand in homework.

Now write the step-by-step sequence to explain where rocks eroded by either wind, moving ice, or water go, using pages C12 and C13 in your book.

____

____

____

____

____

____

Name ______________________________ Date ______________

Lesson Outline
Lesson 1

Earth's Changing Crust

Fill in the blanks.  Reading Skill: **Sequence of Events** - questions 6, 7, 13, 14, 17, 21, 22

What Makes the Crust Move?

1. Cracks in Earth's crust are called ________________.
2. Devices called ________________ record the motions of Earth's crust.
3. Surveyors leave ________________ that tell the exact elevation of a place.
4. Scientists who study Earth are called ________________.
5. The crust is only ________________ of Earth's thickness.
6. Under the crust is the ________________, Earth's thickest layer.
7. Below the mantle is Earth's ________________, with liquid and solid parts.
8. Pieces of Earth's crust that move along the surface are called ________________.

What Forces Act on the Crust?

9. The force that stretches or pulls apart the crust is ________________.
10. The force that squeezes or pushes the crust together is ________________.
11. The force that twists, tears, or pushes one part of the crust past another is ________________.
12. Mountains made of crumpled and folded layers of rock are called ________________.

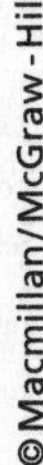

Name ______________________ Date ______________

Lesson Outline
Lesson 1

13. Hot, molten rock deep below Earth's surface is ______________; when it comes to the surface it is ______________.

14. Blocks of crust moving along a fault can form ______________ ______________.

What Other Forces Shape Earth's Surface?

15. The breaking down of materials of Earth's crust into smaller pieces is ______________.

16. The picking up and carrying away of the pieces is called ______________.

17. Changes in temperature cause rock to ______________ and ______________.

How Can Wind and Ice Erode Rock?

18. Wind mostly erodes pieces of rock that are the size of ______________ particles or smaller.

19. A moving river of ice that carries rocks and wears away the land is called a ______________.

Where Do Eroded Rocks Go?

20. The dropping off of bits of eroded rock is called ______________.

21. Deposition can gradually fill up depressions, or ______________, in Earth's surface.

What Forces Shape the Moon's Surface?

22. The only weathering and erosion on the Moon is caused by ______________ that strike its surface.

Name________________________________ Date______________

Interpret Illustrations
Lesson 1

What Makes the Crust Move?

A cross section shows what is inside something. Some cross sections, like this one, show not only what is inside but also what the parts are doing. In this diagram, you should be able to pick out the parts of Earth as well as words that show motion.

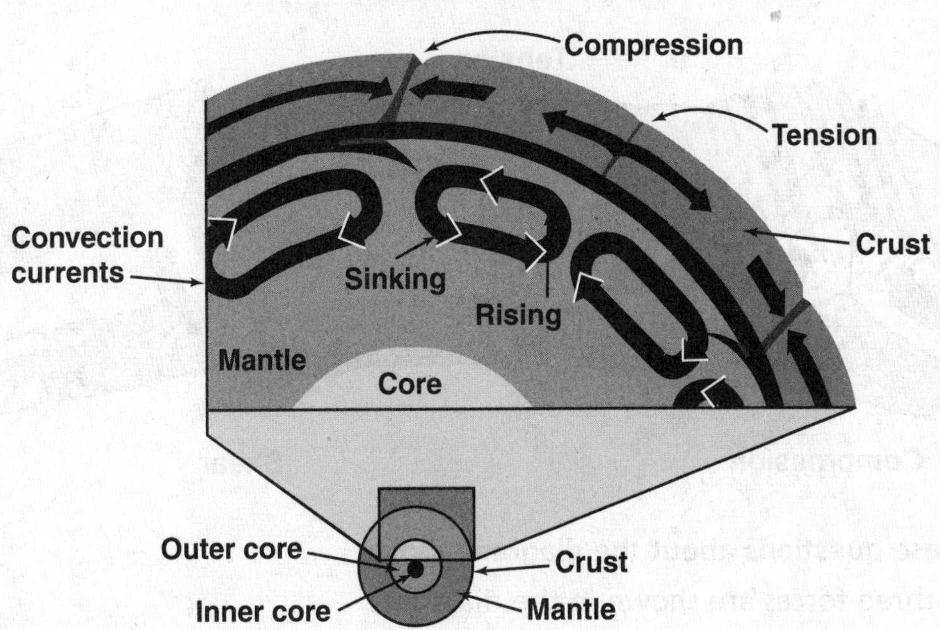

Answer these questions about the diagram above.

1. What three parts of Earth are shown in the diagram?

____

2. What words of motion can you find?

____

3. What is happening during compression?

____

4. How would you describe tension?

____

5. What happens in a convection current?

____

Name ______________________ Date ____________

Interpret Illustrations

Lesson 1

What Forces Act on the Crust?

Diagrams like this one show how forces act. This diagram shows three forces that act on Earth's crust. The arrows tell you the direction in which each force acts and how the force makes the rock layers move.

FORCES IN THE CRUST

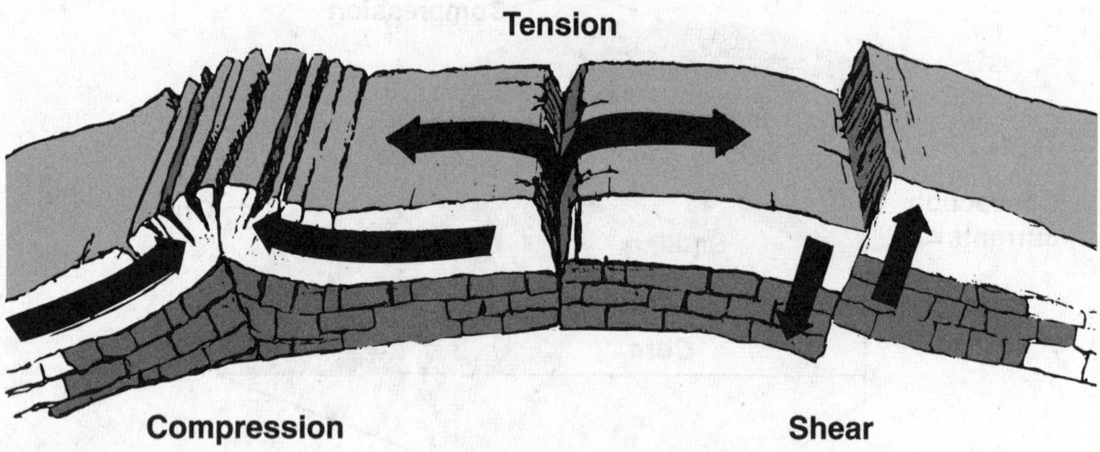

Answer these questions about the diagram above.

1. What three forces are shown in the diagram?

____

2. Which forces act mostly in a horizontal direction?

____

3. Which force acts mostly in a vertical direction? ____________
4. What results from tension between rock layers?

____

5. What happens when a shear force acts on rock layers?

____

6. What does compression do to rock layers?

____

7. Which forces do you think could cause mountains to form?

____

Name ______________________ Date ____________

Lesson Vocabulary
Lesson 1

Earth's Changing Crust

Match the correct letter with the description.

______ 1. magma that reaches Earth's surface

______ 2. the dropping off of bits of eroded rock

______ 3. a crack in the crust, whose sides show evidence of motion

______ 4. the picking up and carrying away of pieces of rock

______ 5. someone who measures Earth's elevation and leaves plaques called bench marks

______ 6. a scientist who studies Earth

______ 7. a delicate device that records the motion of Earth's crust

______ 8. hot, molten rock deep below Earth's surface

______ 9. a chunk of rock from space that strikes a surface, such as Earth

______ 10. the breaking down of rocks into smaller pieces

Vocabulary

- **a. fault**
- **b. geologist**
- **c. magma**
- **d. lava**
- **e. weathering**
- **f. erosion**
- **g. deposition**
- **h. meteorite**
- **i. seismograph**
- **j. surveyor**

Fill in the correct term.

11. stretches or pulls apart the crust: ______________

12. twists, tears, or pushes one part of the crust past another: ______________

13. squeezes or pushes the crust together: ______________

14. mountains made of crumpled and folded layers of rock: ______________

15. mountains formed by blocks of crust moving along a fault: ______________

Name ______________________ Date ______________

Cloze Test

Lesson 1

Earth's Changing Crust

Vocabulary

geologists	crack	seismographs	marks
sea	California	vibrations	predict

Fill in the blanks.

Earthquakes commonly occur where Earth has a fault or ______________ in its crust. One big fault runs through most of the state of ______________. During an earthquake, ______________ travel through the crust. This motion can be recorded by devices named ______________. Surveyors measure elevation or how high a location is above ______________ level. They use plaques called bench ______________ to detect crust movement. Scientists who study Earth are ______________. They try to ______________ an earthquake by recording tiny movements.

Name ______________________ Date __________

Lesson Outline
Lesson 2

Landforms

Fill in the blanks.  Reading Skill: **Sequence of Events** – questions 10, 13, 15

How Do Rivers Change the Land?

1. One of the most important causes of change on Earth is ______________ water.
2. Rivers begin high in mountains or hills as small ______________.
3. ______________ is water that runs off Earth's solid surface.
4. An area from which water is drained is called a(n) ______________.
5. The force of ______________ keeps water flowing downhill.
6. Pieces of material carried by moving water are called ______________.
7. The force of running water with its load of sediment can ______________ a stream bed.
8. ______________ are bends or S-shapes in rivers.
9. Some of the world's most important agricultural areas are found in ______________.
10. Over time, sediments build up, creating ______________ farmlands.
11. The place where a river empties into an ocean is called the ______________ of the river.
12. A fan-shaped deposit of sediment is called a(n) ______________.

How Do Water Gaps, Canyons, and Valleys Form?

13. Small channels that are deepened and widened by erosion form ______________.
14. Deep V-shaped valleys are often called ______________.
15. Over time, a river that slowly cuts its way across and down into resistant rock can form a(n) ______________.

Name ______________________ Date ____________

Lesson Outline

Lesson 2

How Do Beaches, Dunes, and Landslides Form?

16. Water, gravity, wind, waves, and glaciers are all agents of ____________.

17. ____________ form when sediments are deposited on shorelines.

18. Dunes are created when ____________ picks up sand particles and carries them until obstacles, such as sand and pieces of shell, slow the wind speed.

19. Rapid ____________ can be set off by earthquakes, volcanic activity, and heavy rains.

How Do You Read Topographic Maps?

20. A ____________ connects points of equal height above or below sea level.

21. Contour lines that are closer together indicate a steeper ____________.

22. Bodies of water, such as oceans, rivers, and lakes, are indicated on ____________.

What Are Earth's Major Layers?

23. The hard outer layer of Earth is the ____________.

24. Earth's water layer is the ____________.

25. The layers of gases that surround Earth is its ____________.

Name______________________ Date__________

Interpret Illustrations

Lesson 2

How Do Rivers Change the Land?

Look at the picture of the stream below. What effect is the stream having on the mountain?

Answer these questions about the picture above.

1. Where do you think the stream began?

____

____

2. What force keeps the water moving downhill?

____

3. What would cause the stream to move faster?

____

____

4. How do you think the stream is changing the mountain?

____

____

Name______________________________ Date____________

Interpret Illustrations

Lesson 2

What Are Earth's Major Layers?

Look at the illustration below. It shows the three major layers of Earth. What are the important features of each layer?

Answer these questions about the diagram above.

1. How would you describe the atmosphere?

____

____

2. What important gas is contained in the atmosphere?

____

3. The rocky surface that makes up the top layer of the lithosphere is called the crust. Which part of the illustration shows the crust?

____

4. Does the hydrosphere only include Earth's oceans? Explain.

____

____

5. If you looked at a picture of Earth taken from space, would you see more of the lithosphere or the hydrosphere? Explain.

____

____

Name ______________________ Date __________

Lesson Vocabulary
Lesson 2

Landforms

Match the correct letter with the description.

Vocabulary

a. runoff
b. watershed
c. sediment
d. meander
e. flood plain
f. delta
g. lithosphere
h. hydrosphere
i. atmosphere

_______ **1.** a fan-shaped deposit of sediment

_______ **2.** the water layer that covers most of the lithosphere

_______ **3.** areas from which water is drained

_______ **4.** pieces of material carried by water

_______ **5.** many layers of gases that surround Earth

_______ **6.** an area that forms when eroded material from the outer side of a river is deposited on the inner side

_______ **7.** water that runs off Earth's surface

_______ **8.** the hard, outer layer of Earth

_______ **9.** an S-shaped bend in a river

Which of Earth's major layers—lithosphere, hydrosphere, or atmosphere—is linked to each phrase.

10. includes fresh water found in lakes, rivers, streams, groundwater, and ice

____

11. made up of four major layers ______________________

12. is about 100 km (62.14 mi) thick ______________________

13. outer layers help protect Earth from forms of harmful energy from the Sun

____

14. contains trillions of liters of water ______________________

15. has a layer called the troposphere ______________________

Name ______________________ Date __________

Cloze Test

Lesson 2

Landforms

Vocabulary

runoff	watersheds	sediment	meanders	flood plains
delta	lithosphere	hydrosphere	atmosphere	

Fill in the blanks.

Tributaries are fed by ______________. The areas from which water is drained are called ______________. Moving water carries ______________, or pieces of material, with it. ______________ are bends or S-shapes in rivers. ______________ may form along the banks of a river as the landscape flattens. A fan-shaped deposit of sediment is called a(n) ______________. The ______________ is the hard, outer layer of Earth. Trillions of liters of water are found in the ______________. The ______________ is made up of many layers of gases that surround Earth.

Name ____________________ Date __________

Lesson Outline
Lesson 3

Minerals of Earth's Crust

Fill in the blanks.  Reading Skill: **Sequence of Events** - questions 7, 11, 13, 14

How Can You Identify a Mineral?

1. Solid materials of Earth's crust are called ____________.
2. Like all matter, minerals are made of ____________.
3. Most minerals are chemical ____________, that is, two or more elements joined together.
4. Minerals form geometric shapes, called ____________.
5. You should always observe ____________ on a fresh surface.
6. The way light bounces off a mineral is called ____________.

How Else Can You Identify Minerals?

7. The color of the powder left when a mineral is rubbed against a hard, rough surface is called its ____________.
8. The measure of how well a mineral resists scratching is called its ____________.
9. If a mineral breaks along flat surfaces, it is called ____________.
10. Minerals that do not break smoothly are said to have ____________.

How Do Minerals Form?

11. Many rocks form when hot liquid rock, or ________________, cools.
12. Some of the rarest minerals, like ________________, form deep within Earth.
13. Cooling hot water can form ________________ that settle to the bottom of the water.
14. Common table salt is formed when ocean water ________________.

What Are Minerals Used For?

15. A mineral that contains a useful substance is called a(n) ________________.
16. Iron comes from the mineral ________________.
17. Substances that conduct electricity and can be stretched into wires are ________________.
18. Minerals valued for being rare and beautiful are called ________________.

Name ______________________ Date ____________

Interpret Illustrations

Lesson 3

How Can You Identify a Mineral?

These pictures illustrate six minerals of Earth's crust. The diagrams show that each mineral forms a different geometric shape. To compare the diagrams of the shapes, look at the lengths of the sides of each crystal. Also notice the angles at which those sides meet.

The mineral chalcopyrite (kal•kuh•PIGH•right) is a compound made of the elements copper, iron, and sulfur. It is where much of our copper comes from. Copper is used for wire, coins, pots, and pans.

Tetragonal crystal

Triclinic crystal

The mineral kaolinite (KAY•uh•luh•night) is used in china plates and ceramic objects. It comes in many colors–red, white, reddish brown, and even black.

Rock salt, which is used to melt ice, is the mineral halite (HAL•ight). It is a compound made of the elements sodium and chlorine.

Cubic crystal

Monoclinic crystal

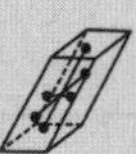

The "lead" in a lead pencil is not the metal element lead at all. It is the mineral graphite (GRAF•ight), which is a form of the element carbon.

Hexagonal crystal

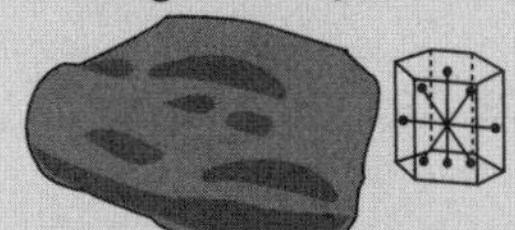

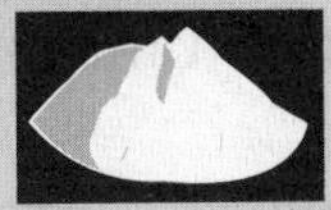

Topaz is a mineral used in many kinds of jewelry. It comes in many colors– pink, pale blue, and even yellow or white.

Orthorhombic crystal

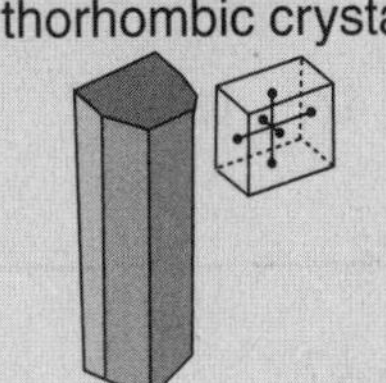

Talc is the mineral used in talcum powder. Talc comes in white and greenish colors.

Answer these questions about the diagram above.

1. From what mineral do we get most of our copper? ____________
2. What is graphite? ____________
3. What is rock salt? ____________
4. What is topaz? ____________
5. Which crystals have sides that tilt? ____________
6. Which crystal has the most sides? ____________
7. What is one difference between a tetragonal crystal and a cubic crystal?

____

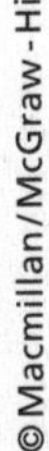

Name________________________ Date______________

Interpret Illustrations

Lesson 3

How Else Can You Identify Minerals?

This table shows calcite and its properties. Read about each of the properties of calcite.

MINERAL	COLOR(S)	LUSTER (Shiny as Metals)	PORCELAIN PLATE TEST (Streak)	CLEAVAGE (Number)	HARDNESS (Tools scratched by)	DENSITY (Compared with water)
Calcite	colorless, white, pale blue	no	colorless, white	yes—3	3 (all but fingernail)	2.7

Answer these questions about the table above.

1. Why is the color of calcite listed as colorless, white, or pale blue?

____

2. What does it mean that calcite has no luster?

____

3. What would be a good title for this table?

____

4. How would you test for hardness?

____

____

____

5. What does a density of 2.7 mean?

____

Name ______________________ Date ____________

Lesson Vocabulary

Lesson 3

Minerals of Earth' s Crust

You are given an unknown mineral. How would you identify it? Fill in the table below.

Property	Definition	How Accurate Is It? Explain.
color	1.	2.
luster	3.	4.
hardness	5.	6.
streak	7.	8.
cleavage	9.	10.

Give an example of each and describe a property.

1. crystals ______________________
2. ores ______________________
3. gems ______________________
4. metals ______________________

Answer the questions.

5. Define mineral and describe why minerals are a nonrenewable resource.

____

____

Name ______________________ Date ____________

Cloze Test
Lesson 3

Minerals of Earth's Crust

Vocabulary

plate	luster	fracture	cleavage
color	scratching	metallic	diamond

Fill in the blanks.

You can identify minerals by the ______________ or way the light bounces off their surfaces. Minerals with a(n) ______________ luster are shiny. Another way to tell minerals apart is by observing ______________ on a fresh surface. You can tell gold from pyrite by using a streak ______________. Hardness, or how well a mineral resists ______________, is measured by Mohs' scale. The hardest mineral is a(n) ______________. If a mineral has ______________, it will break along flat surfaces. If it has ______________, it wouldn't break smoothly.

Name ____________________ Date ____________

Lesson Outline

Lesson 4

Earth's Rocks and Soil

Fill in the blanks. Reading Skill: **Sequence of Events** - questions 2, 6, 10, 11

How Are Rocks Alike and Different?

1. A ____________________ is any naturally formed solid in the crust made up of one or more kinds of minerals.

What Are Igneous Rocks?

2. Rocks that form when melted rock material cools and hardens are called ____________________ rocks.

What Are Sedimentary Rocks?

3. Small bits of matter joined together form ____________________ rock.

How Are Sedimentary Rocks Useful?

4. Sandstone and limestone are both used to make ____________________.

5. Sedimentary rocks sometimes contain ____________________, or the remains or imprints of living things of the past.

What Are Metamorphic Rocks?

6. A rock formed under heat and pressure from another kind of rock is called a(n) ____________________ rock.

How Are Metamorphic Rocks Used?

7. A metamorphic rock that contains minerals with brilliant colors and is easy to carve is ____________________.

Name ______________________ Date ____________

Lesson Outline

Lesson 4

Where Does Soil Come From?

8. The main ingredient in soil is weathered ______________.
9. Decayed plant and animal material in soil is called ______________.

How Can People Ruin Soil?

10. People bury garbage and ______________ in soil and spray ______________ on soil to kill pests.
11. People also throw foam cups and plastic materials on the ground, which contributes to ______________.

How Can People Protect the Soil?

12. People can add materials called ______________ and humus to replace minerals removed by crops.
13. Farmers can plow furrows across a slope rather than up and down, which is called ______________.

What Is the Rock Cycle?

14. The never-ending series of processes by which rocks change from one form to another is called the ______________.

Name ______________________________ Date ____________

Interpret Illustrations
Lesson 4

How Are Metamorphic Rocks Used?

This diagram shows how coal is formed. Notice that above the drawings there is a bar that tells how long ago each step took place. The oldest stage is at the left, and the present stage is shown at the right.

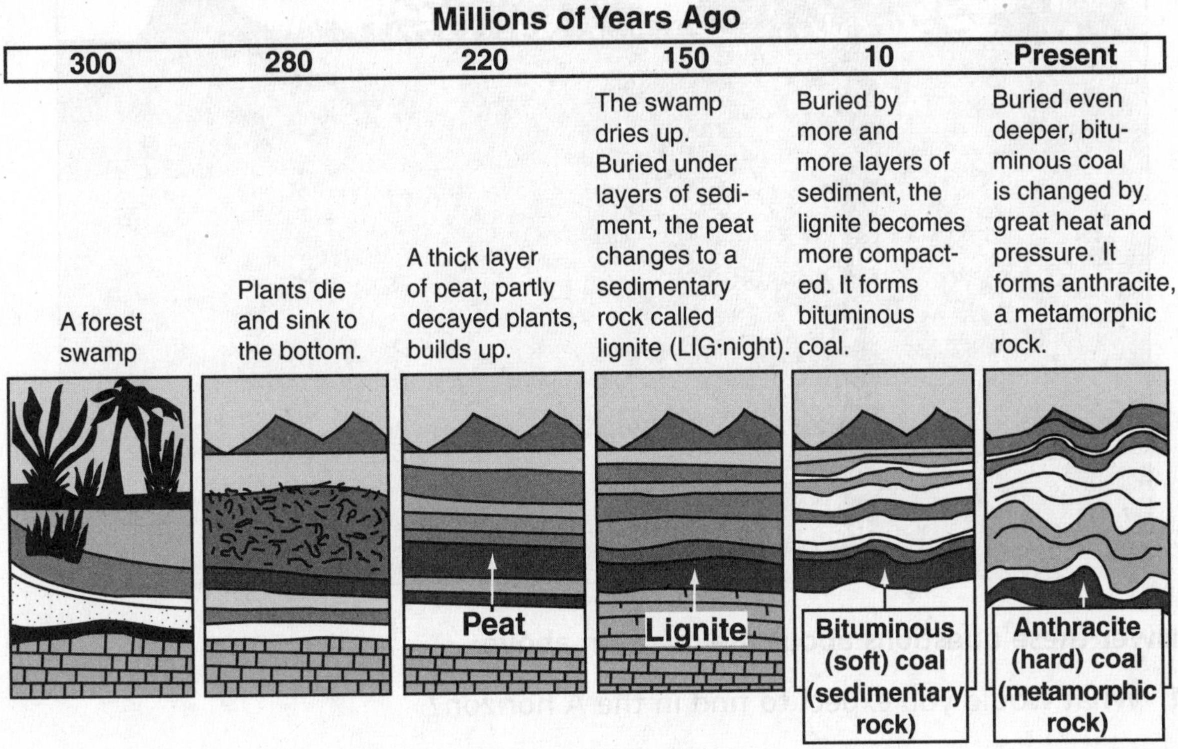

Millions of Years Ago

300	280	220	150	10	Present
A forest swamp	Plants die and sink to the bottom.	A thick layer of peat, partly decayed plants, builds up.	The swamp dries up. Buried under layers of sediment, the peat changes to a sedimentary rock called lignite (LIG·night).	Buried by more and more layers of sediment, the lignite becomes more compacted. It forms bituminous coal.	Buried even deeper, bituminous coal is changed by great heat and pressure. It forms anthracite, a metamorphic rock.
		Peat	Lignite	Bituminous (soft) coal (sedimentary rock)	Anthracite (hard) coal (metamorphic rock)

Answer these question about the diagram above.

1. Where does coal come from? ______________________________

2. What is peat? ______________________________

3. What does peat change into? ______________________________

4. What happens to the lignite? ______________________________

____

5. What does lignite become? ______________________________

6. How is bituminous coal changed to anthracite coal? ______________________________

____

7. What kind of rock is anthracite coal? ______________________________

Name ______________________ Date ____________

Interpret Illustrations
Lesson 4

Where Does Soil Come From?

This cross section shows what you would see if you could slice through Earth's crust. You would see layers, or horizons, of soil. The arrows point to the top and bottom of each of the three horizons, and to the top of the bedrock layer.

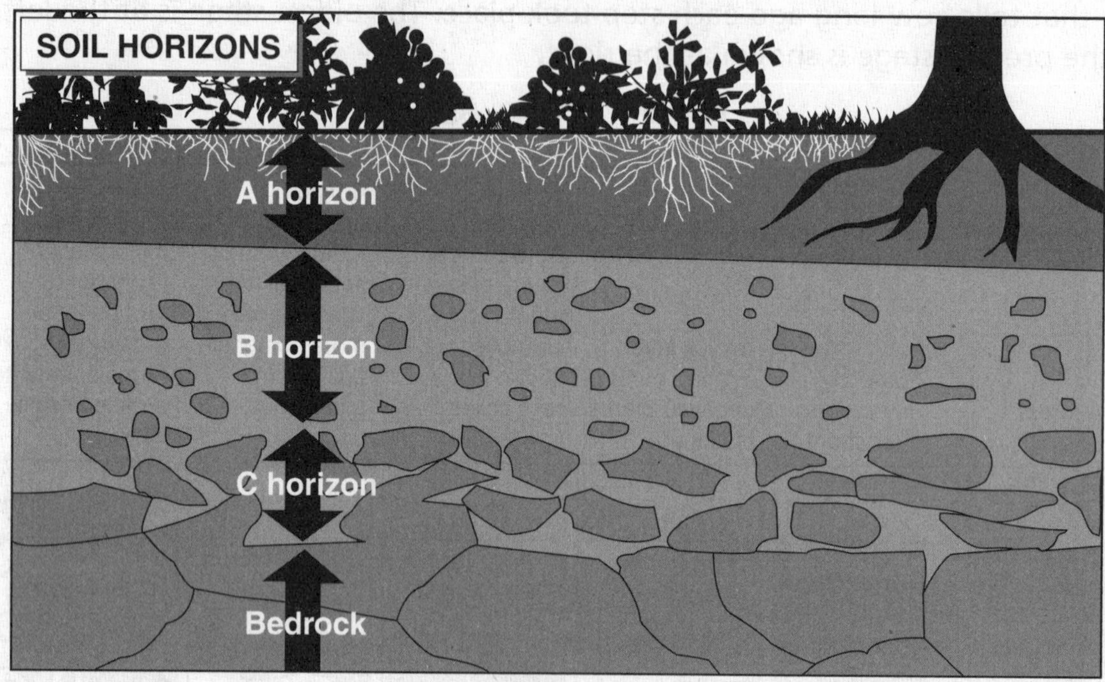

Answer these questions about the diagram above.

1. What would you expect to find in the A horizon?

____

2. What is in the B horizon? ______________________

3. How do the rocks of the C horizon differ from those in the B horizon?

____

4. Into which layer or layers do the roots of most plants reach? ____________

5. How do you think the bedrock got its name?

____

6. Which rocks do you think are most like the bedrock? Why? ____________

____

Name________________________________ Date____________

Earth's Rocks and Soil

Match the correct letter with the description.

______ 1. rocks changing from one form into another in a never-ending series of processes

______ 2. decayed plant or animal material in soil

______ 3. any remains or imprint of living things of the past

______ 4. a naturally formed solid in the crust, made up of one or more minerals

______ 5. adding any harmful substances to Earth's land, water, or air

______ 6. a rock formed from another kind of rock under heat and pressure

______ 7. a rock made of bits of matter joined together

______ 8. a rock formed when melted rock material cools and hardens

Vocabulary

a. rock
b. fossil
c. humus
d. pollution
e. rock cycle
f. igneous rock
g. sedimentary rock
h. metamorphic rock

Answer the question.

9. What does soil have to do with rocks?

____

____

Name ______________________ Date ____________

Cloze Test

Lesson 4

Earth's Rocks and Soil

Vocabulary

chemicals	nutrients	wastes	forest
pollution	decay	graze	erosion

Fill in the blanks.

People ruin soil by burying hazardous ______________ and garbage in it. They spray soil with ______________ to kill weeds. These materials add up to ______________ which is adding any harmful substance to Earth's land, water, or air. Trash like plastic that doesn't ______________ makes soil unusable. When people grow the same crops every year, it depletes the ______________ in soil. If plants are removed, wind and rain will expose soil to ______________. Another cause is allowing cattle to ______________ too long in the same spot. The soil is exposed whenever a(n) ______________ is cut down.

Name________________________ Date____________

Chapter Vocabulary

Chapter 7

Landforms, Rocks, and Minerals

Circle the letter of the best answer.

1. Rivers flowing along slopes carry pieces of material called

a. meteorites. **b.** deltas.

c. sediments. **d.** meanders.

2. Important agricultural areas because of their fertile soil are called

a. faults. **b.** deltas.

c. crusts. **d.** ores.

3. Earth's water layer is called the

a. atmosphere. **b.** crust.

c. hydrosphere. **d.** lithosphere.

4. The layer above the lithosphere is the

a. atmosphere. **b.** hydrosphere.

c. bedrock. **d.** soil.

5. Iron and aluminum are useful metals made from

a. shears. **b.** gems.

c. depositions. **d.** ores.

6. A crack in Earth's crust with sides that show evidence of motion is a(n)

a. cleavage. **b.** deposition.

c. fault. **d.** shear.

7. A scientist who studies Earth is called a(n)

a. geologist. **b.** astronomer.

c. mineralogist. **d.** gemologist.

Name_________________________ Date_____________

Chapter Vocabulary

Chapter 7

Circle the letter of the best answer.

8. Hot, molten rock deep below Earth's surface is
   a. deposition.
   b. humus.
   c. lava.
   d. magma.

9. The breaking down of rocks into smaller pieces is called
   a. deposition.
   b. erosion.
   c. pollution.
   d. weathering.

10. A chunk of rock from space that strikes Earth is a(n)
   a. gem.
   b. meteorite.
   c. mineral.
   d. ore.

11. A solid material of Earth's crust with a definite composition is a(n)
   a. fossil.
   b. mineral.
   c. volcano.
   d. meteorite.

12. How well a mineral resists scratching is called its
   a. cleavage.
   b. hardness.
   c. luster.
   d. streak.

13. A rock formed when melted rock material cools and hardens is
   a. humus.
   b. igneous.
   c. metamorphic.
   d. sedimentary.

14. Any remains or imprint of living things of the past describes a(n)
   a. fossil.
   b. gem.
   c. ore.
   d. rock.

Chapter Summary

1. What is the name of the chapter you just finished reading?

2. What are four vocabulary words you learned in the chapter? Write a definition for each.

3. What are two main ideas that you learned in this chapter?

Name______________________________ Date______________

Chapter Graphic Organizer

Chapter 8

Air, Water, and Energy

Each energy source listed in the word bank below is either a fossil fuel or an alternative energy source. Also, each energy source is either a renewable resource or a nonrenewable resource. Write each energy source in the correct box at the bottom of the page. After you have classified each energy source, underline any that pollute the environment and circle any that come from biomass.

For example, propane is a fossil fuel. It is also a nonrenewable source of energy. Propane has been written in the correct box. It is not underlined or circled, because it does not come from biomass, and it does not pollute the environment when it is burned.

Word Bank

animal wastes	**coal**	**falling water**
gasoline	**geothermal energy**	**natural gas**
oil	**plant matter**	**propane**
sunlight	**wind**	**wood**

	Renewable Resource	**Nonrenewable Resource**
Fossil Fuel		propane
Alternative Energy Source		

Name ______________________ Date ____________

Chapter Reading Skill

Chapter 8

Draw Conclusions

Read each description. Then draw and label your conclusion of what it is.

Water goes around with me! I mean, I renew, or recycle, it so you can have fresh water to drink. What am I? ____________	The wind spins my "arms" around. That energy runs generators that make electricity. What am I? ____________
I carry energy from place to place. Each water particle in me moves in a circle. When I break I can be very powerful. What am I? ____________	People say I hang like a brown cloud over many cities. I'm a mixture of smoke and fog. What am I? ____________

Name________________________________ Date____________

Chapter Reading Skill
Chapter 8

Check the Facts

Remember, you can't draw a conclusion unless you have all the facts. Read each set of facts below. Can you draw a conclusion? If not, circle **not enough facts.** If you can, circle your answer.

1. Fossil fuels formed as the remains of ancient plants and animals decayed. Dinosaurs were ancient animals. Are some fossil fuels formed from the decayed remains of dinosaurs?

   yes **no** **not enough facts**

2. Wastes from burning fossil fuels go into the air. They can mix with moisture to produce acid rain. Should people stop burning fossil fuels forever?

   yes **no** **not enough facts**

3. Steve's father was a football star for his high school and for his college. Steve plans to go to the same high school and college his dad did. Will Steve be a football star in those schools?

   yes **no** **not enough facts**

4. The fifth graders at Sloane School started a recycling program. The kids have also looked for ways to save electricity in the school and in their homes. Next Saturday they plan to clean up a nearby park and plant some flowers and shrubs. Do the fifth graders care about our environment?

   yes **no** **not enough facts**

5. Kyle is shorter than Mark but taller than Alex. Alex is shorter than Rudy. Is Rudy taller than Kyle?

   yes **no** **not enough facts**

6. Marta got a 94 on the test. Luis and Diane got 92. Danny, Jackie, and Allie got 90, and Roger had 89. The other kids had lower scores. Did Marta do better on the test than the other kids?

   yes **no** **not enough facts**

Name ______________________ Date ____________

Lesson Outline
Lesson 5

Earth's Atmosphere

Fill in the blanks.  Reading Skill: **Draw Conclusions** - question 7

Why Do Living Things Need the Atmosphere?

1. Air is a mixture of nitrogen, ________________, and water vapor.
2. On land, living things have structures that enable them to get ________________ directly from air.
3. Oxygen helps provide us with energy by breaking down ________________.
4. After food is broken down, living things give off the gas ________________.
5. The gas that is given off by animals is used by ________________ to make food.
6. The process by which plants make food and give off oxygen is called ________________.
7. Oxygen is called a(n) ________________ resource because it can be replaced.
8. The ________________ protects us from the Sun's harmful ultraviolet rays.
9. The atmosphere also protects us from rocks from space. The ________________ with the air burns them up before they can come to Earth.
10. Clouds in the atmosphere help regulate ________________.
11. About 78 percent of the air is ________________, which is an important part of proteins.

Name ______________________________ Date ______________

Lesson Outline
Lesson 5

What Causes Pollution?

12. Many pollutants get into the air from burning ______________.
13. These fuels were formed from the decay of ancient ______________.
14. Fossil fuels are ______________ resources.
15. Other fossil fuels include ______________, ______________, and ______________.
16. Even plowing farm fields can cause air pollution by adding ______________ to the atmosphere.
17. Natural events like ______________ and ______________ can add particles and smoke to the air.
18. Many cities experience ______________, a mixture of smoke and fog.
19. At ground level, the gas ______________ can make people sick.

What Is Acid Rain?

20. Burning fossil fuels can cause harmful chemicals called acids to form in the atmosphere, which then fall to Earth in the form of ______________.

How Can We Clean Up the Air?

21. The United States passed the ______________ in 1967 to decrease air pollution.
22. To protect the air, we can limit the amount of ______________ in refrigerators and air conditioners.

Name ________________________ Date ____________

Interpret Illustrations

Lesson 5

Why Do Living Things Need the Atmosphere?

This diagram shows the carbon dioxide-oxygen cycle. Notice how the flow is circular. By following the arrows you can see how carbon dioxide and oxygen circulate through the atmosphere.

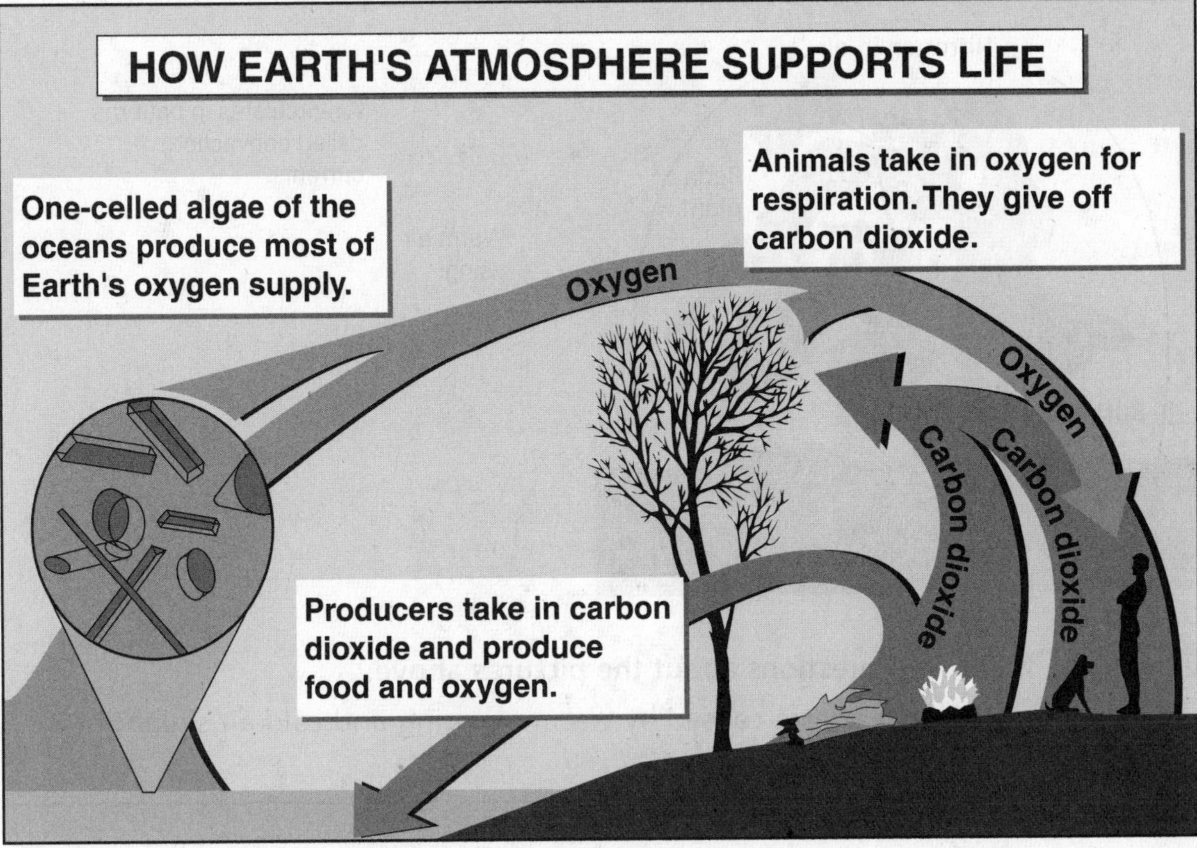

HOW EARTH'S ATMOSPHERE SUPPORTS LIFE
One-celled algae of the oceans produce most of Earth's oxygen supply.
Animals take in oxygen for respiration. They give off carbon dioxide.
Oxygen
Oxygen
Carbon dioxide
Carbon dioxide
Producers take in carbon dioxide and produce food and oxygen.

Answer these questions about the diagram above.

1. What role do trees play in the carbon dioxide-oxygen cycle?

____

2. What organisms produce most of Earth's oxygen supply?

____

3. What role do animals and humans play in this cycle?

____

4. What do producers produce besides oxygen? ____________________

Name ____________________ Date ____________

Interpret Illustrations

Lesson 5

How the Atmosphere Supports Life

The diagrams below show two ways that the atmosphere helps to protect life on Earth. The first picture shows how nitrogen is cycled. The second picture shows how convection currents circulate air temperatures.

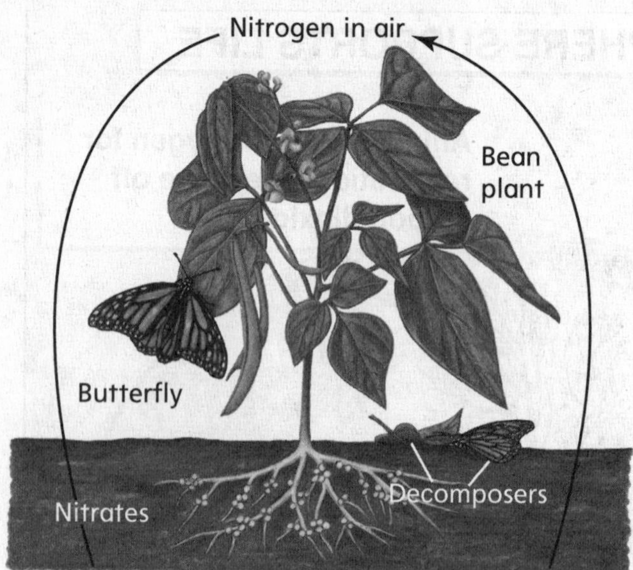

Air circulates in patterns called convection currents.

Warm air rising

Cold air sinking

Answer the following questions about the pictures above.

1. What kind of currents are caused by warm air rising and cold air sinking?

____

____

2. When is nitrogen returned to the soil?

____

____

3. What do plants use to make proteins?

____

____

Name ____________________ Date __________

Lesson Vocabulary
Lesson 5

Earth's Atmosphere

Match the correct letter with the description.

_____ 1. a mixture of smoke and fog

_____ 2. a resource that can be replaced in a short period of time

_____ 3. a resource that cannot be replaced in your lifetime

_____ 4. a fuel formed from the decay of ancient forms of life

_____ 5. moisture that falls to Earth after being mixed with wastes from burned fossil fuels

_____ 6. the part of the atmosphere that screens out much of the Sun's UV rays

Vocabulary

a. renewable resource

b. ozone layer

c. fossil fuel

d. smog

e. acid rain

f. nonrenewable resource

Write a one-sentence summary to answer each of these questions.

7. How do animals use oxygen? ____________________

8. How does air protect living things? ____________________

9. What makes air dirty? ____________________

10. How can rain be harmful? ____________________

11. How can we clean up the air? ____________________

12. How does smog affect you? ____________________

Name ______________________ Date ____________

Cloze Test

Lesson 5

Earth's Atmosphere

Vocabulary

ozone	fossil fuels	CFCs	breathing
fires	smog	radiation	

Fill in the blanks.

Many pollutants get into the air from burning ________________. Natural events such as grass and forest ________________ spread smoke. A mixture of smoke and fog known as ________________ hangs over many cities like a brown cloud. It can kill people who have ________________ problems. There are holes in the ________________ layer. These allow harmful UV ________________ through. Substances called ________________ are released from products like refrigerators and aerosol spray cans. They rise into the atmosphere where they can affect the ozone layer.

Name ______________________ Date ____________

Lesson Outline

Lesson 6

Earth's Fresh Water

Fill in the blanks.  Reading Skill: **Draw Conclusions** - questions 5, 7, 9, 11, 15

How Do We Use Earth's Oceans?

1. The oceans provide us with food we call ______________ and minerals.
2. Nodules, or lumps, of ______________ can be picked up from the sea floor.
3. Offshore rigs pump ______________ and natural gas from beneath the ocean floor.
4. The process by which we get fresh water from seawater is called ______________.
5. Water leaves dissolved materials behind when it ______________.

Where Is Fresh Water From?

6. Fresh water is constantly renewed in the ______________.
7. Water is on the move–as a liquid that changes to a ______________ and then back to a ______________.
8. The main source of water in the water cycle is the ______________.
9. The main sources of fresh water on land are ______________ and ______________, which fall as ______________.
10. Plants give off ______________.

Name ______________________ Date ______________

11. The water that seeps into the spaces between bits of rock and soil is called ______________.

12. The top of the water-filled spaces is called the ______________.

13. An underground layer of rock or soil that is filled with water is called a(n) ______________.

14. Some groundwater seeps out of the ground in a(n) ______________.

15. A hole dug below the water table so water can be brought up for drinking is called a(n) ______________.

How Can Fresh Water Be Polluted?

16. Rain and snow can pick up pollutants in the ______________.

17. Runoff water can carry pollutants from ______________ into streams or lakes.

18. As water soaks down through the soil, it can pick up chemicals called ______________.

How Can We Purify Water?

19. The ground acts as a fine screen to trap, or ______________, dirt particles in water.

20. In water purification plants, water is filtered by ______________.

21. To kill bacteria, ______________ is added to water.

Name ______________________ Date ____________

Interpret Illustrations
Lesson 6

Where Is Fresh Water From?

This diagram shows the water cycle, a giant circle in which water moves around Earth and the atmosphere. Follow the circle from number 1 to number 5. Also notice the arrows that show the movement of water into and through Earth.

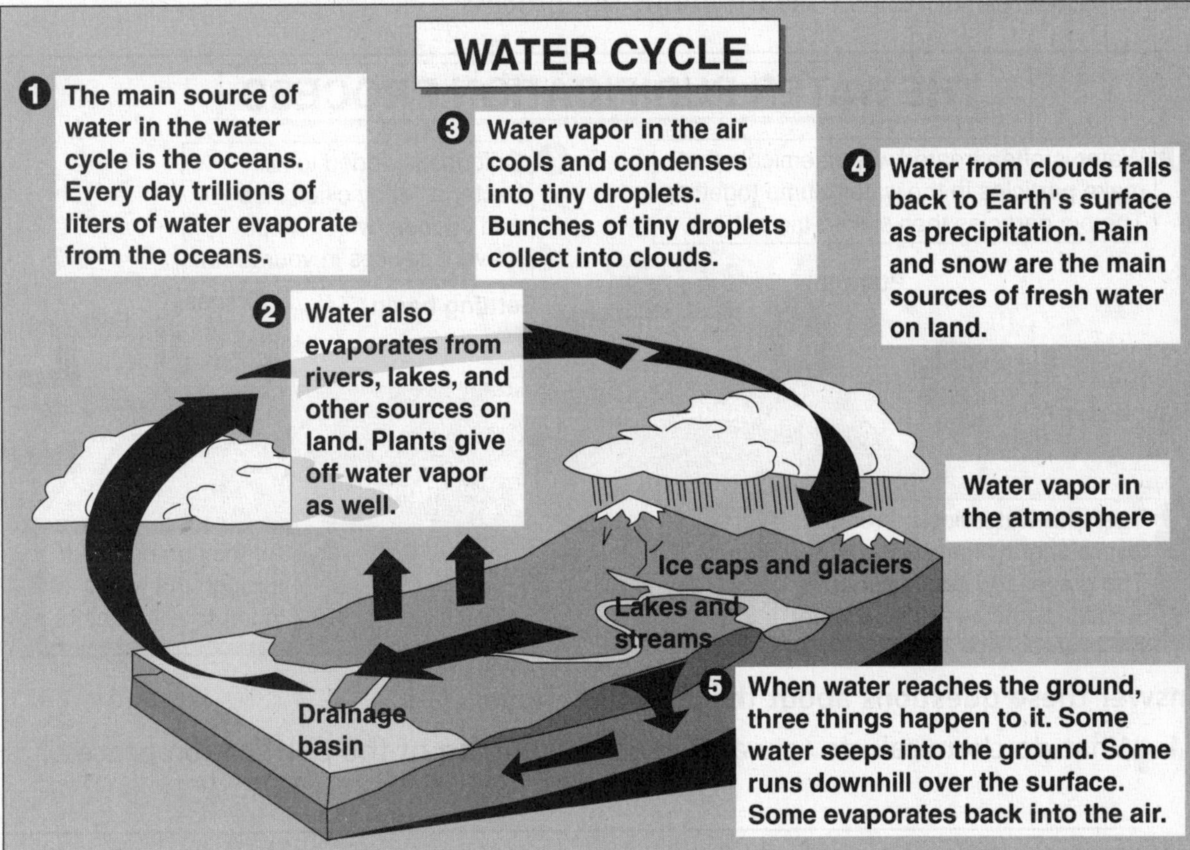

WATER CYCLE
1 The main source of water in the water cycle is the oceans. Every day trillions of liters of water evaporate from the oceans.
3 Water vapor in the air cools and condenses into tiny droplets. Bunches of tiny droplets collect into clouds.
4 Water from clouds falls back to Earth's surface as precipitation. Rain and snow are the main sources of fresh water on land.
2 Water also evaporates from rivers, lakes, and other sources on land. Plants give off water vapor as well.
Water vapor in the atmosphere
Ice caps and glaciers
Lakes and streams
Drainage basin
5 When water reaches the ground, three things happen to it. Some water seeps into the ground. Some runs downhill over the surface. Some evaporates back into the air.

Answer these questions about the diagram above.

1. Where does most of the water in the water cycle come from? ____________
2. What are other sources of water? ____________
3. How are rain droplets formed? ____________
4. What are the main sources of fresh water on land? ____________
5. What happens after water reaches the ground? ____________

____

____

6. What process moves water into the air? ____________
7. What process brings water down to Earth? ____________

Name ______________________ Date __________

Interpret Illustrations

Lesson 6

How Can We Purify Water?

This diagram shows how water is purified so we can drink it. Although individual water purification plants vary, the basic elements are the same. Follow the arrows to see the direction in which water moves through the process. Identify each step and read the paragraphs to learn about the process.

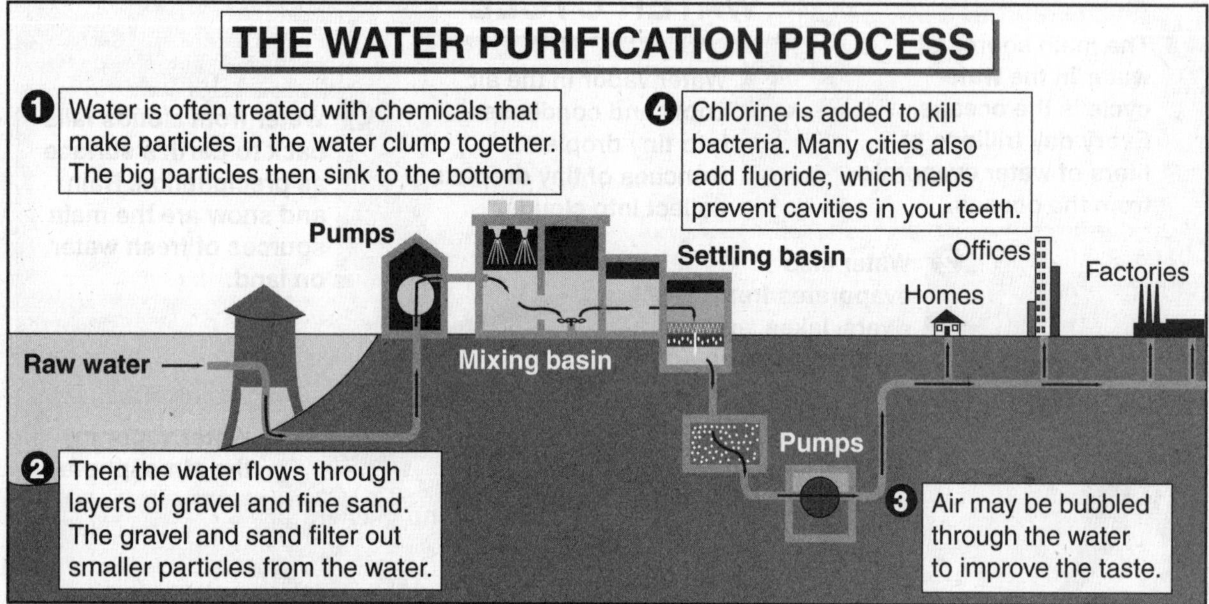

THE WATER PURIFICATION PROCESS
1 Water is often treated with chemicals that make particles in the water clump together. The big particles then sink to the bottom.
4 Chlorine is added to kill bacteria. Many cities also add fluoride, which helps prevent cavities in your teeth.
Pumps
Settling basin
Offices
Factories
Homes
Raw water
Mixing basin
Pumps
2 Then the water flows through layers of gravel and fine sand. The gravel and sand filter out smaller particles from the water.
3 Air may be bubbled through the water to improve the taste.

Answer these questions about the diagram above.

1. What do chemicals do to water at the beginning of the purification process?

____

____

2. What do fine sand and gravel do to the water?

____

3. What is used to improve the taste of water? ______________

4. What chemical is added to kill bacteria? ______________

5. How do you benefit when fluoride is added to drinking water?

____

6. What is used to move the water along during the purification process? ______

____

Name____________________________ Date______________

Lesson Vocabulary
Lesson 6

Earth's Fresh Water

Match the correct letter with the description.

Vocabulary

a. desalination
b. water cycle
c. groundwater
d. water table
e. aquifer
f. spring
g. well
h. reservoir
i. salt water
j. ocean

______ 1. from this we get seafood, minerals, and fossil fuels

______ 2. water that seeps into the ground into spaces between bits of rock and soil

______ 3. a place where groundwater seeps out of the ground

______ 4. the continuous movement of water between Earth's surface and the air; it helps clean water

______ 5. an underground layer of rock or soil filled with water

______ 6. a storage area for freshwater supplies

______ 7. a hole dug below the water table that water seeps into

______ 8. getting freshwater from seawater

______ 9. most of Earth's water is this type

______ 10. the top of the water-filled spaces in the ground

Write "ocean water" or "freshwater" to identify if these things are more likely to pollute ocean water or freshwater.

11. water traveling over garbage ______________________

12. a barge going out to dump wastes ______________________

13. acid rain ______________________

14. spills from oil tankers ______________________

15. pesticides ______________________

Name ______________________ Date ____________

Cloze Test
Lesson 6

Earth's Fresh Water

Vocabulary

desalination	wells	spring	seawater
reservoirs	pipelines	table	aquifers

Fill in the blanks.

You can't use ________________ to drink or sprinkle plants. A process called ________________ allows you to obtain freshwater from oceans. People in cities get water from ________________, storage areas for freshwater. This water is transported by ________________ to locations where it's needed. Some people tap into groundwater by digging ________________. These holes are dug beneath the water ________________. Other holes are dug into ________________ sandwiched between rock layers. Some groundwater seeps out in a place called a(n) ________________.

Name ______________________ Date __________

Lesson Outline
Lesson 7

Earth's Oceans

Fill in the blanks.  Reading Skill: **Draw Conclusions** - questions 1, 3, 4, 6, 10, 12, 16

What Are Oceans, Seas, and Basins?

1. Most of Earth's water is contained in large bodies of salt water called ______________.
2. The ocean floor, or ______________, contains mountains, valleys, and plains.
3. If all the water evaporated from the oceans, a layer of ______________ about 60 meters (200 feet) thick would be left on the ocean floor.
4. When volcanoes erupt, they let out water ______________ and other gases.
5. Seawater is an important source of ______________.
6. Drilling rigs are used to extract ______________ and ______________ from the land beneath the sea.

What Causes Ocean Currents?

7. A stream of water that flows through the ocean like a river is a ______________.
8. ______________ currents are driven by the wind.

How Does the Water in a Wave Move?

9. As a wave passes through the ocean, water ______________ move in circles.
10. Each ______________ may hurl thousands of tons of water against the shore, breaking rocks apart and smoothing the fragments into pebbles and sand.

Name ______________________ Date __________

Lesson Outline

Lesson 7

What Causes the Tides?

11. The pull of the Moon and Sun's gravity on Earth causes ______________.
12. The combined pull of the Sun and the Moon during times of New and Full Moons causes the highest high tides and lowest low tides, called ______________ tides.
13. ______________ tides have the smallest range.

What Is the Ocean Floor Like?

14. The underwater edge of a continent is called its ______________.
15. The ______________ leads steeply down from the continental shelf toward the sea floor.
16. The ______________ is a buildup of sediment on the sea floor at the bottom of the continental slope.
17. One of the flattest places on Earth, the ______________ plain, is at the end of the continental rise.
18. A huge underwater mountain is called a ______________.
19. A V-shaped valley below the ocean is known as a ______________.
20. The mountain range rising above the ocean floor of the Atlantic Ocean is known as the ______________.

How Do We Explore the Oceans?

21. Scientists explore the ocean with a ______________ map by tracing the patterns of sonar echoes.
22. ______________ are formed when seawater trickles down into the hot, newly formed oceanic crust saturated with minerals.

How Do People Affect the Oceans?

23. Offshore drilling for oil and ______________ harms the environment.

Name _______________ Date _______

Interpret Illustrations

Lesson 7

What Causes Ocean Currents?

The map below shows the major ocean currents of the world. Read each label and note the solid and broken arrows.

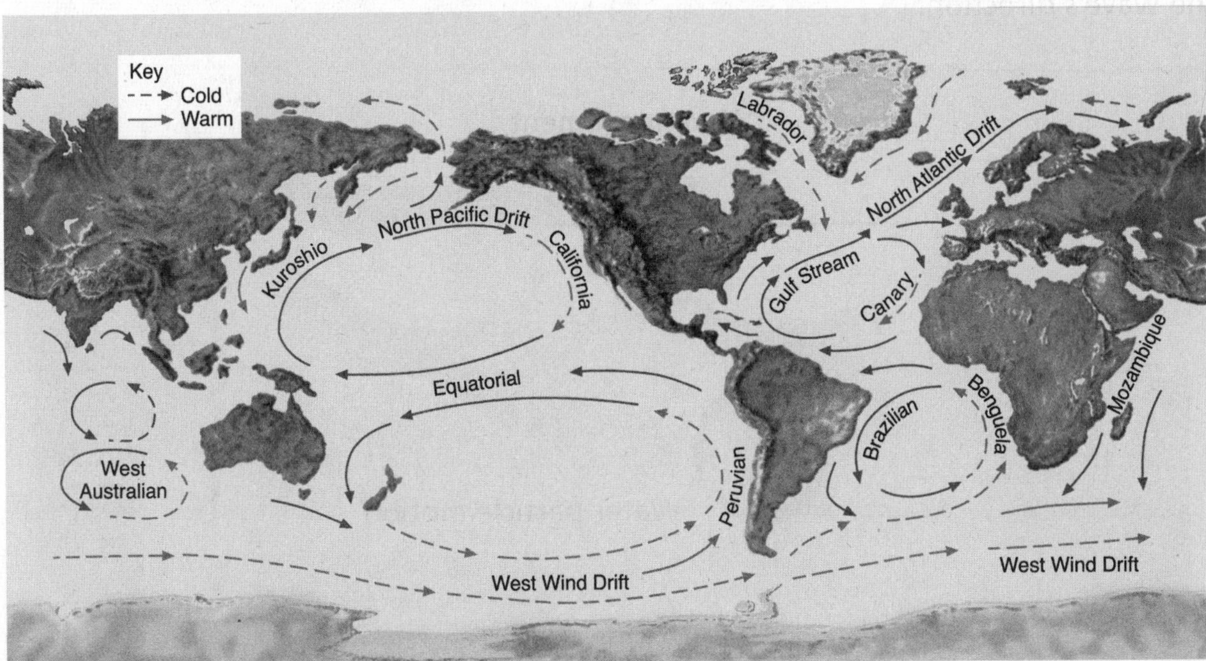

Answer the following questions about the map above.

1. What is the name of the most important warm current flowing along the Atlantic coast of North America?

____

____

2. What effect does the California Current have on the West Coast of the United States?

____

____

3. What is the name of the most important warm current that flows along the Atlantic coast of South America?

____

____

Name________________________________ Date______________

Interpret Illustrations

Lesson 7

How Does the Water in a Wave Move?

The illustration below shows the movement of a wave, water particle motion, and the wave's direction.

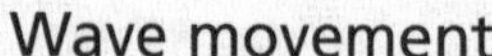

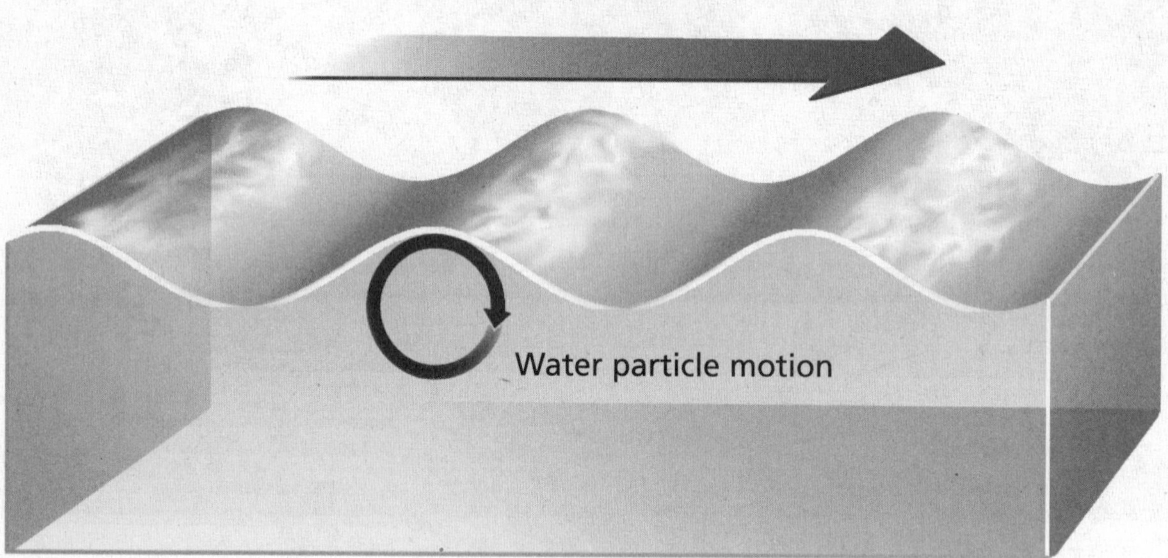

Answer the following questions about the picture above.

1. In what direction does the water particle move as the wave passes over it?

____

____

2. What happens to each water particle as a wave passes through the ocean?

____

____

3. What brings energy to the ocean water?

____

____

Name ______________________ Date __________

Lesson Vocabulary

Lesson 7

Earth's Oceans

Vocabulary

- current
- basin
- abyssal plain
- continental shelf
- continental rise
- seamount
- trench
- mid-ocean ridges
- continental slope

Fill in the blanks.

1. The ocean's ______________, or floor, contains mountains, valleys, and plains.
2. The ______________ leads from the continental shelf toward the sea floor.
3. A stream of water that flows through the ocean like a river is called a(n) ______________.
4. A buildup of sediment on the sea floor at the bottom of the continental slope is called the ______________.
5. A huge underwater mountain is called a(n) ______________.
6. The underwater edge of a continent is the ______________.
7. A narrow v-shaped valley on the ocean floor is known as a ______________.
8. The ______________ are chains of mountains that wind their way through all the world's major oceans.
9. Vast, flat land covering almost half of the deep ocean floor is called the ______________.

Answer each question.

10. What percentage of fish species are being overfished? ______________
11. Where does marine pollution come from?

____

____

Name ______________________ Date ______________

Cloze Test
Lesson 7

Earth's Oceans

Vocabulary

circles	currents	denser	Surface
Deep-water	meters	resurface	bend

Fill in the blanks.

Earth's rotation affects surface ______________. This causes currents to ______________ to the right in the Northern Hemisphere and to the left in the Southern Hemisphere. The currents start flowing in huge ______________. ______________ currents move at a speed of about 220 kilometers a day. ______________ currents are set in motion by differences in the temperature and saltiness of water. Colder, saltier water is ______________ than the water below it and slowly sinks toward the ocean bottom. The water in a deep-water current moves much slower than the water in a surface current, just a few ______________ a day. The water may not ______________ for another 500 years.

Name ____________________ Date __________

Lesson Outline
Lesson 8

Energy Resources

Fill in the blanks.  Reading Skill: **Draw Conclusions** - questions 7, 14, 18

How Are Fossil Fuels Turned into Energy?

1. Energy to heat homes and run many common devices such as lights, computers, radios, TVs, and washers comes from ____________.
2. Electricity from power plants comes to homes through ____________.
3. Most vehicles get their energy from burning ____________.
4. Heat from fossil fuels is used to boil water and turn it into ____________ in order to generate electricity.
5. When the steam is released, it is directed to a big, pinwheel-like ____________.
6. The spinning turbine turns a ____________ to make electricity.
7. There are two disadvantages to fossil fuels: they are ____________, meaning they cannot be replaced, and they ____________ the environment.

Where Do Fossil Fuels Come From?

8. Fossil fuels are the remains of ____________.
9. The source of energy in fossil fuels can be traced back to the ____________.
10. The world's largest consumer of energy is ____________.

Name ______________________________ Date ______________

Lesson Outline

Lesson 8

Fill in the blanks.

What Other Sources of Energy Are There?

11. Sources of energy other than the burning of fossil fuels are called ____________________ energy sources.

12. Hydroelectric plants use running or falling ____________________ to spin a ____________________ to make electricity.

13. Earth's internal heat, or ________________________, can be used to heat homes and produce electricity.

14. Plants harness the Sun's energy through the process of ____________________.

15. A house can be heated when sunlight is trapped by ____________________ ____________________, or collectors.

16. Devices that convert sunlight into electric energy are called ____________________.

How Can We Conserve Energy?

17. Conserving energy means not ____________________ what we have and using as ____________________ as possible.

18. We can also use ______________________________ like water, wind, and solar energy.

19. Renewable energy from plant matter and animal wastes is called ____________________.

Name________________________________ Date____________

Interpret Illustrations

Lesson 8

How Are Fossil Fuels Turned into Energy?

This diagram shows how most homes get their energy. It shows a picture of a power plant, power lines, a transformer, and a home.

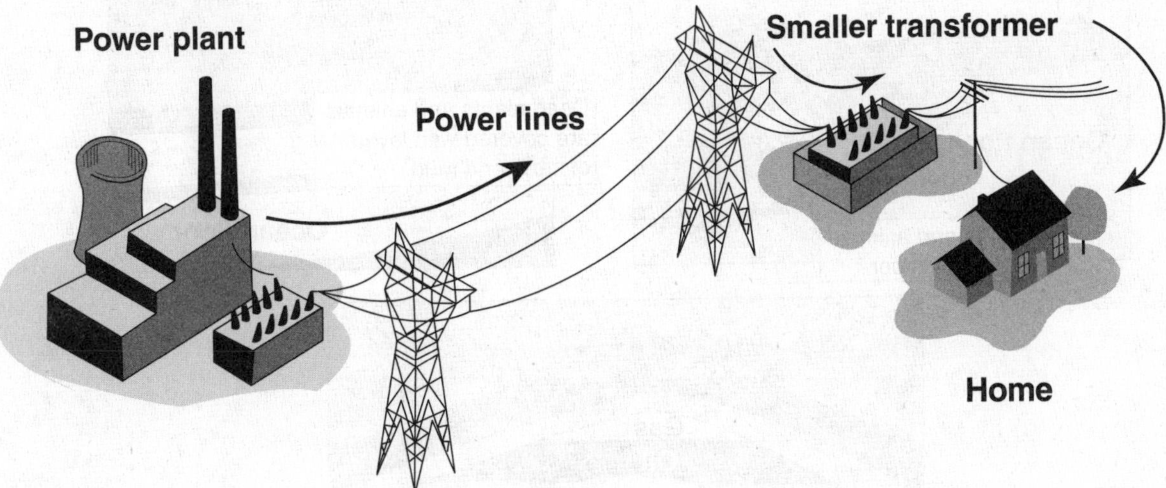

Answer these questions about the diagram above.

1. Where do homes get their energy?

____

2. How does energy get from the power plant to a house?

____

3. How do the wires change after the electricity goes through the smaller transformer?

____

____

Name____________________ Date__________

Interpret Illustrations

Lesson 8

Where Do Fossil Fuels Come From?

This diagram shows the process of microscopic plants and animals forming gas and oil over millions of years. Be sure you identify what happened and what caused it to happen.

HOW FOSSIL FUELS ARE FORMED

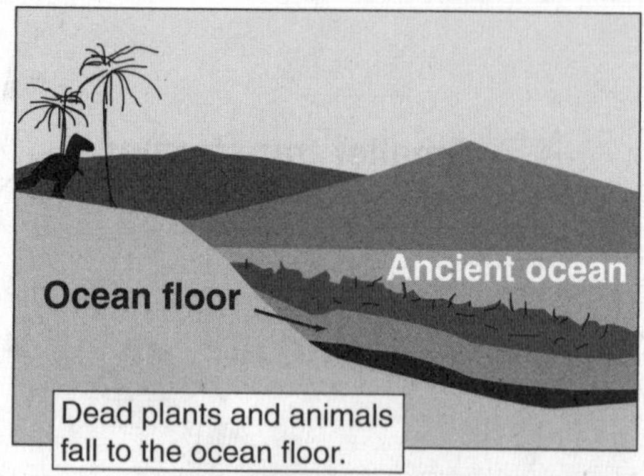

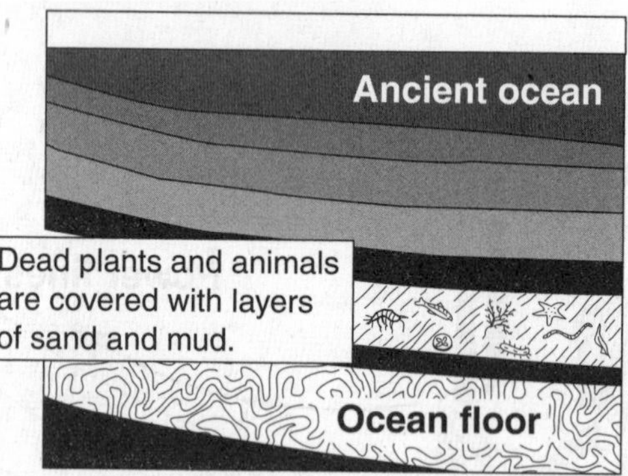

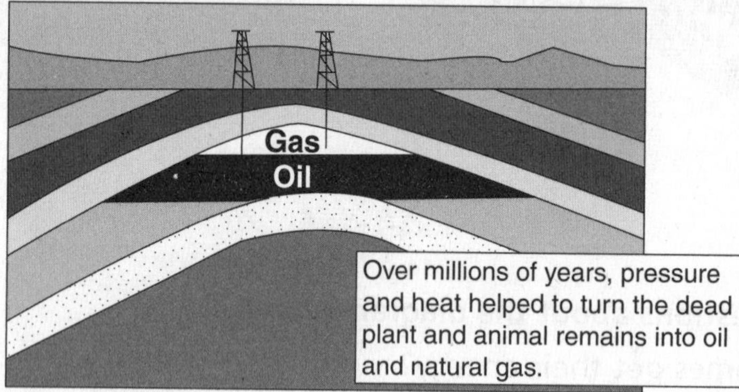

Gas
Oil
Over millions of years, pressure and heat helped to turn the dead plant and animal remains into oil and natural gas.

Answer these questions about the diagram above.

1. Where do oil and natural gas come from?

____

2. Where did this change take place?

____

3. What happened in the ocean after the microscopic plants and animals fell to the floor?

____

4. What forces act over millions of years to produce oil and natural gas?

____

Name ______________________ Date ____________

Lesson Vocabulary

Lesson 8

Energy Resources

Match the correct letter with the description.

_____ **1.** energy from plant matter or animal wastes

_____ **2.** called an inexhaustible source of energy

_____ **3.** Earth's internal heat

_____ **4.** uses moving water to produce electricity

_____ **5.** energy from the Sun

_____ **6.** coal, oil, and natural gas that come from plants and animals that died long ago

_____ **7.** caused by burning fossil fuels

_____ **8.** a source of energy other than the burning of a fossil fuel

Vocabulary

a. alternative energy source
b. geothermal energy
c. biomass
d. fossil fuels
e. air pollution
f. Sun
g. hydroelectric plant
h. solar energy

Answer each question.

9. List three ways you are using energy right now.

____

____

____

10. List three ways you can conserve energy.

____

____

Name ______________________ Date ____________

Cloze Test
Lesson 8

Energy Resources

Vocabulary

conserve	wind	alternative	generator	renewable
sunlight	steam	biomass	geothermal energy	

Fill in the blanks.

Fossil fuel supplies are limited since they aren't a(n) ________________ energy source. We need to ________________ energy supplies so they're not exhausted. One ________________ energy source is the waterwheel. It can be used in a hydroelectric plant, where water spins a(n) ________________. Moving air or ________________ also can be harnessed for electricity. The ________________ from water heated by ________________ produces electricity. Solar cells convert ________________ into electric energy. Another form of energy comes from ________________, or plant and animal wastes.

Name________________________ Date____________

Chapter Vocabulary

Chapter 8

Air, Water, and Energy

Circle the letter of the best answer.

1. A resource that can be replaced in a short period of time is

a. alternative. **b.** fossil.

c. nonrenewable. **d.** renewable.

2. A layer of gas in the atmosphere that screens out much of the Sun's UV rays is called the

a. air cycle. **b.** aquifer.

c. ozone layer. **d.** reservoir.

3. A fuel formed from the decay of ancient forms of life is a(n)

a. alternative fuel. **b.** biomass.

c. fossil fuel. **d.** geothermal fuel.

4. A mixture of smoke and fog is called

a. acid rain. **b.** biomass.

c. ozone. **d.** smog.

5. Moisture that falls to Earth after being mixed with wastes from burned fossil fuels is

a. acid rain. **b.** groundwater.

c. biomass. **d.** the water cycle.

6. Getting fresh water from seawater is

a. conservation. **b.** deposition.

c. desalination. **d.** purification.

7. The continuous movement of water between Earth's surface and the air is the

a. geothermal cycle. **b.** ozone layer.

c. water cycle. **d.** water table.

Name ______________________ Date ____________

Chapter Vocabulary

Chapter 8

Circle the letter of the best answer.

8. Water that seeps into the ground into spaces between bits of rock and soil is
   a. groundwater.
   b. reservoir water.
   c. spring water.
   d. well water.

9. The top of the water-filled spaces in the ground is the
   a. ozone layer.
   b. reservoir.
   c. water cycle.
   d. water table.

10. An underground layer of rock or soil filled with water is the
   a. aquifer.
   b. groundwater.
   c. reservoir.
   d. water table.

11. A place where groundwater seeps out of the ground is a(n)
   a. aquifer.
   b. spring.
   c. water table.
   d. well.

12. A hole dug below the water table that water seeps into is a(n)
   a. aquifer.
   b. reservoir.
   c. spring.
   d. well.

13. A storage area for freshwater supplies is a(n)
   a. aquifer.
   b. reservoir.
   c. spring.
   d. well.

14. Earth's internal heat is called
   a. biomass.
   b. fossil fuel.
   c. geothermal energy.
   d. hydroelectric power.

Name ______________________ Date ____________

Unit Vocabulary

Unit C

Use Words

Word Box					
weathering	luster	lithosphere	smog	pollution	lava
hydrosphere	geologist	trench	igneous	humus	magma

Read each question below. Write your answer on the line. Use a word from the box.

1. Which word identifies melted rock deep below Earth's surface? ____________
2. Which word means "the way light bounces off a mineral's surface"?

   ____
3. Which word identifies a scientist who studies Earth? ____________
4. Which word identifies all of Earth's water? ____________
5. Which word identifies magma that reaches Earth's surface? ____________
6. Which word describes rocks breaking down into smaller pieces?

   ____
7. Which word means "a long, narrow V-shaped valley in the ocean"?

   ____
8. Which word describes a mixture of smoke and fog? ____________
9. Which word identifies a rock that forms when melted rock material cools and hardens?

   ____
10. Which word means the decayed plant or animal material in soil?

    ____
11. Which word describes the hard outer layer of Earth? ____________
12. Which word describes harmful substances added to Earth's land, water, or air?

    ____

Name ______________________ Date ____________

Unit Vocabulary

Unit C

Scrambled Words

Read each clue, then unscramble the letters and write the word correctly. Use the Word Box to check your spelling.

Word Box				
desalination	**aquifer**	**fossil**	**lava**	**magma**
crust	**cleavage**	**geologist**	**reservoir**	**meteorite**

1. melted underground rock — **gamma** ____________
2. a layer above Earth's mantle — **struc** ____________
3. magma that flows from a volcano — **aavl** ____________
4. a tendency of a mineral to break along flat surfaces — **cvgealea** ____________
5. imprint or remains of living things of the past — **silsof** ____________
6. someone who studies Earth — **togglesoi** ____________
7. storage area for freshwater supplies — **rivoreser** ____________
8. rock from space that strikes Earth — **trimoteee** ____________
9. underground rock filled with water — **fraquie** ____________
10. getting freshwater from seawater — **nationalside** ____________

Now mix up the letters of four more words from this unit. Write a clue for each word, then ask a friend to unscramble it.

1. ____________________________
2. ____________________________
3. ____________________________
4. ____________________________

Name______________________________ Date____________

Word Webs

Remember that a word web lists words that describe or relate to the same thing. Here is a web for the word *lava*.

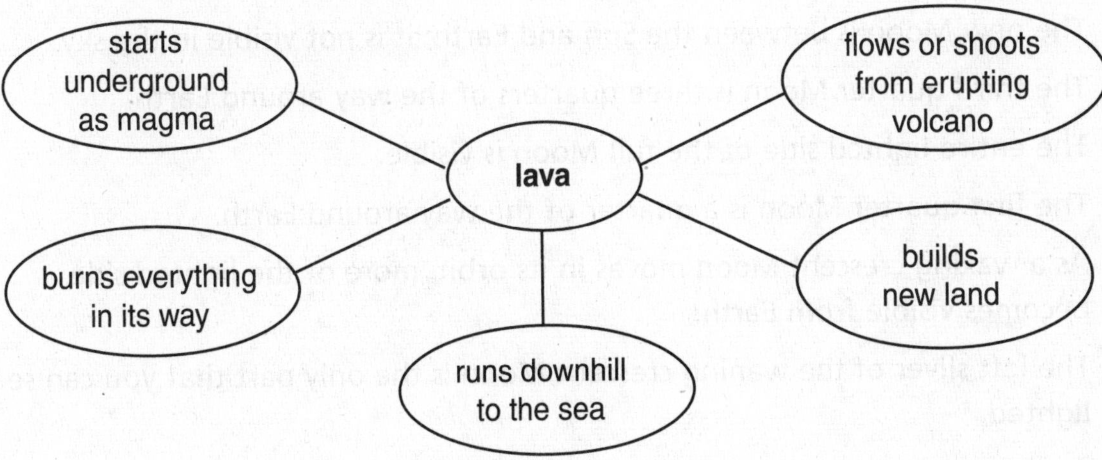

Now choose a vocabulary word or phrase, such as *biomass, solar energy, pollution, resource, or seamount.* Make a web with other words that relate to or describe it.

Name ______________________ Date ____________

Chapter Graphic Organizer

Chapter 9

Astronomy

Listed below are 8 phases of the Moon. Put the phases in the proper sequence in the diagram. Two of the phases have been completed for you. Write the number of each phase in the correct circle and the name of each phase on the line. Then shade in how each phase would appear.

1. The new Moon is between the Sun and Earth; it is not visible in the sky.
2. The third quarter Moon is three quarters of the way around Earth.
3. The entire lighted side of the full Moon is visible.
4. The first quarter Moon is a quarter of the way around Earth.
5. As a waxing crescent Moon moves in its orbit, more of the lighted side becomes visible from Earth.
6. The left sliver of the waning crescent Moon is the only part that you can see lighted.
7. As the waning gibbous Moon continues to move in its orbit, less of the lighted side is visible from Earth.
8. The waxing gibbous Moon is almost full.

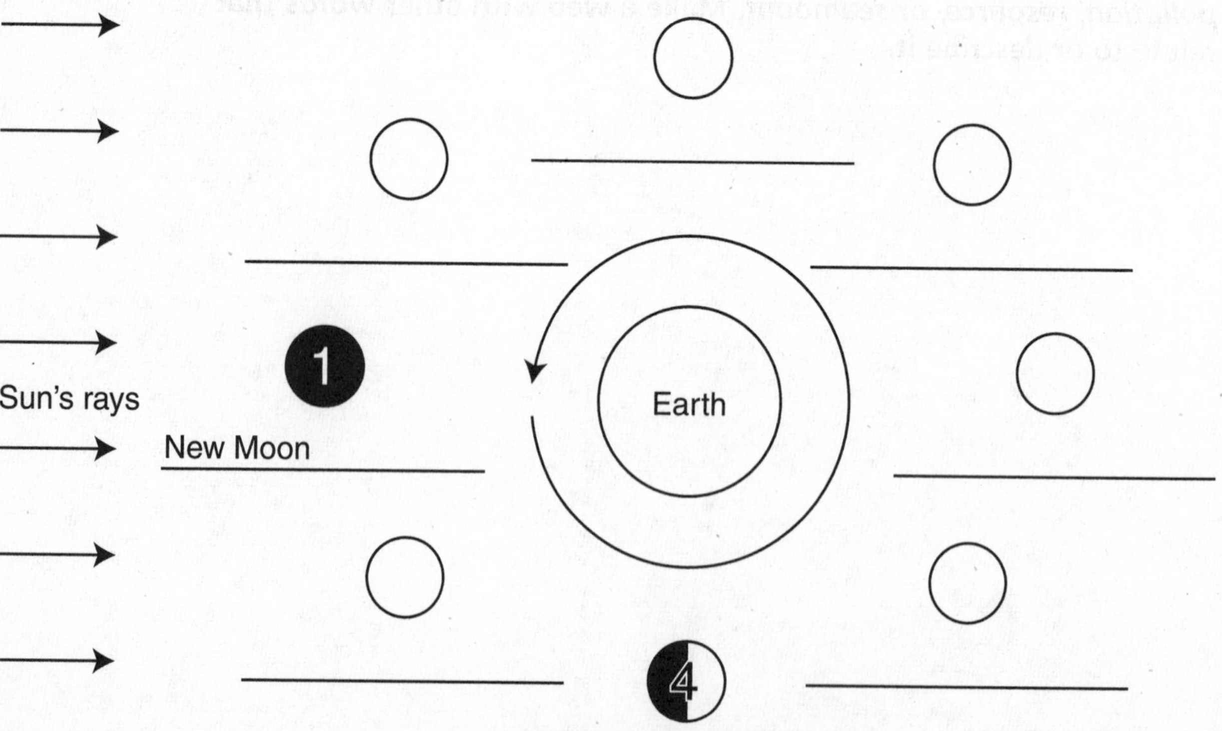

Name ______________________ Date ____________

Sequence of Events

It takes 365 days for Earth to orbit the Sun. Some planets take a shorter time while other planets take longer. A chart is a good way to compare a sequence of events.

Look at the chart below. Then answer the questions.

Planet	Average Distance to the Sun (million km)	Time It Takes (in Earth days) to Orbit the Sun
Mercury	57.9	88 days
Venus	108.2	225 days
Earth	149.6	365 days
Mars	227.9	687 days
Jupiter	778.4	4,331 days
Saturn	1,427	10,756 days
Uranus	2,871	30,687 days
Neptune	4,498	60,190 days
Pluto	5,906	90,553 days

1. Which planet takes about twice as much time as Uranus to orbit the Sun?

____

2. Which planet takes almost 2 Earth years to orbit the Sun?

____

3. After Pluto, which planet takes the greatest amount of time to orbit the Sun?

____

4. After Mercury, which planet takes the least amount of time to orbit the Sun?

____

5. How does distance affect the time it takes a planet to orbit the Sun?

____

____

Name______________________________ Date____________

Chapter Reading Skill

Chapter 9

Follow the Sequence

In our solar system, each planet has a different length of day. The table below lists the planets according to their distance from the Sun. Mercury is the closest to the Sun, and Pluto is the furthest away from the Sun.

Look at the table. Then answer the sequence of events questions below.

Length of Day

Planet	Day = time for complete spin (in Earth hours or days)
Mercury	59 days
Venus	243 days
Earth	24 hours
Mars	24 hours 37 minutes
Jupiter	9 hours 56 minutes
Saturn	10 hours 40 minutes
Uranus	17 hours 14 minutes
Neptune	16 hours 7 minutes
Pluto	6.39 days

1. Which planet has a day length that is almost a year long?

____

2. Which planet has a day length similar to Earth?

____

3. Which planet has a day length about a week long?

____

4. Identify the planets that have shorter days than Earth. Put them in order from shortest to longest lengths.

____

____

Name ______________________ Date ____________

Lesson Outline

Lesson 1

Earth and Its Neighbors

Fill in the blanks.  Reading Skill: **Sequence of Events** - questions 4, 5

What Is the Solar System?

1. The ______________ is the Sun and the objects traveling around it.
2. The objects around the Sun include nine ______________.
3. The path of a planet is called its ______________.
4. Earth is the ______________ planet from the Sun.
5. A complete trip of Earth around the Sun takes one ______________.

What Keeps the Planets in Orbit?

6. The force of attraction between the Sun and Earth is called ______________.
7. Gravity depends on two measurements: ______________.
8. The closer two objects are, the ______________ the pull of gravity between them.
9. Newton described gravity as a property of all ______________.
10. The planets are kept in orbit by the Sun's ______________.
11. The tendency of a moving object to keep moving in a straight line is called ______________.
12. Both ______________ and ______________ keep the planets in their orbits.

Name ______________________________ Date ____________

Lesson Outline
Lesson 1

What Makes a Day?

13. All planets spin, or ______________, like huge spinning tops.
14. How much light and warmth a planet receives depends on ______________________________.
15. The length of a day on Earth is ______________.

What Is the Moon Like?

16. The Moon has no hydrosphere; there is no ______________ to drink; there is no ______________.
17. The Moon has a ______________ surface.
18. Dark-colored regions of the Moon are called ______________.
19. Huge dents on the surface of the Moon are called ______________.
20. The Moon changes shape, or ______________, from day to day.
21. The light of the Moon comes from the ______________ rays striking it.
22. It takes the Moon ______________ days to complete all its phases.

What Are Constellations?

23. In the past, people looked at the stars and saw them arranged in groups that formed patterns in the sky; they called these patterns ______________.

Name ____________________ Date __________

Interpret Illustrations

Lesson 1

What Keeps the Planets in Orbit?

This diagram has three arrows to show three different possible paths that Earth could take. The solid arrow shows Earth's actual path. The dashed arrows show what would happen if only gravity or inertia acted on Earth. Compare the three paths.

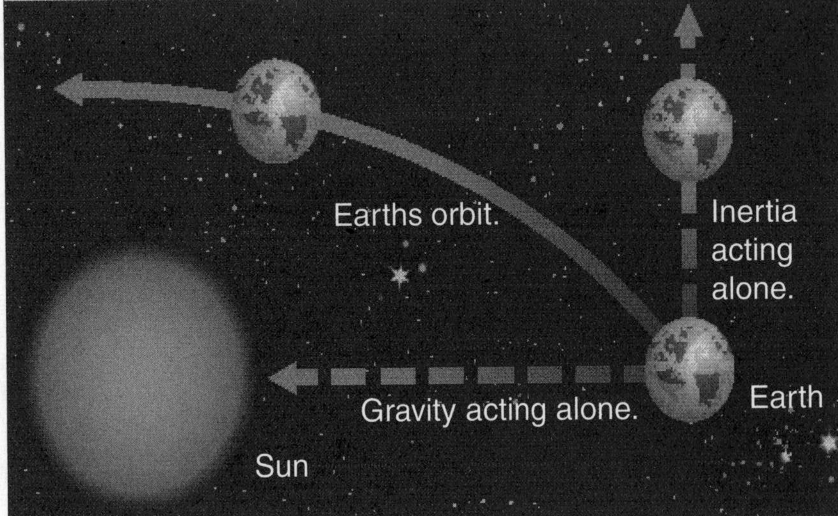

Answer these questions about the diagram above.

1. What would happen if only gravity were operating on Earth?

____

2. Would this path take Earth closer to the Sun or farther away from the Sun?

____

3. What would happen if only inertia were acting on Earth?

____

4. Would this path take Earth closer to the Sun or farther away from the Sun?

____

5. Why does Earth's path bend around the Sun?

____

____

6. What do you think would happen if Earth were not moving?

____

Name ______________________________ Date ______________

Interpret Illustrations

Lesson 1

What Makes a Day?

Look at the table below. It lists the length of day for each of the nine planets in our solar system. Look at the illustration. It shows how Earth rotates and receives sunlight. Study the table and illustration carefully.

Length of Day	
Planet	Day = time for complete spin (in Earth hours or days)
Mercury	59 days
Venus	243 days
Earth	24 hours
Mars	24 hours 37 minutes
Jupiter	9 hours 56 minutes
Saturn	10 hours 40 minutes
Uranus	17 hours 14 minutes
Neptune	16 hours 7 minutes
Pluto	6.39 days

Answer these questions about the table and illustration above.

1. How would you define the length of a planet's day?

____

____

2. Which planet has the shortest day? ______________

3. Which planet's day is closest in length to Earth's day? ______________

4. What is the name of the imaginary line through Earth? ______________

5. In what direction does Earth turn? ______________

Name______________________________ Date______________

Lesson Vocabulary
Lesson 1

Earth and Its Neighbors

Match the correct letter with the description.

Vocabulary

a. solar system
b. planet
c. gravity
d. inertia
e. revolves
f. Earth
g. constellations

_______ **1.** a force of attraction between any object and any other objects around it

_______ **2.** how Earth travels around the Sun

_______ **3.** groups of stars that form patterns in the sky

_______ **4.** the Sun and the objects that are traveling around it

_______ **5.** the third planet from the Sun

_______ **6.** any of the nine objects that travel around the Sun

_______ **7.** the tendency of a moving object to keep moving in a straight line

Fill in the blanks to describe how the Sun affects Earth.

8. The Sun holds Earth in orbit because of _________________ and _________________.

9. The Sun also provides Earth with _________________ and _________________.

10. The Sun's light shines more or less directly on parts of Earth because of Earth's _________________.

Name ______________________ Date ____________

Cloze Test

Lesson 1

Earth and Its Neighbors

Vocabulary

rotates	system	orbit	third
reflect	24	gravity	mass

Fill in the blanks.

The Sun and the objects that travel around it form the solar ____________________. Each planet travels in a path called a(n) ____________________. These planets don't give off light; instead they ____________________ sunlight. Each planet spins, or ____________________, making a complete turn during its day. For Earth, this day-night cycle takes ____________________ hours. Earth is the ____________________ planet from the Sun. An invisible force called ____________________ keeps Earth from flying off into space. This force of attraction depends on distance and ____________________.

Name ______________________ Date ____________

Lesson Outline
Lesson 2

The Solar System

Fill in the blanks.

How Do The Inner Planets Compare?

1. The planets closest to the Sun are the ________________.
2. The inner planets are ________________, Venus, Earth, and Mars.
3. None of the inner planets have ________________.
4. ________________ is the hottest planet.
5. The largest of the inner planets is ________________.

How Do The Outer Planets Compare?

6. The outer planets lie beyond the ________________.
7. Jupiter, Saturn, Uranus, and Neptune are known as the ________________.
8. In comparison, ________________ is tiny and icy.
9. The period of ________________ is much slower among the outer planets.
10. ________________ is less dense than water.
11. A day on ________________ is 17 1/4 hours long.
12. Winds on ________________ can reach speeds of almost 1250 miles an hour.

13. The ____________________ lies between the orbits of Mars and Jupiter.

14. Space rocks that burn up before hitting the ground are ____________________.

15. Space rocks that reach the ground are ____________________.

16. ____________________ are mixtures of ice, rock, and dust.

17. The ____________________ is a region beyond the orbit of Neptune.

Are There Other Solar Systems?

18. Our Sun is only one of billions of stars in the ____________________.

19. The Milky Way is only one of billions of star systems in the ____________________.

20. ____________________ hopes to launch a number of missions to look for Earth-like planets.

21. The first of the missions is called ____________________.

22. The ____________________________ is a space telescope that is planned to launch in 2013.

How Do The Inner Planets Compare?

Look at the illustrations and captions below. Each illustration relates to the caption below it. Read each caption carefully. How are the four planets alike? How are they different.

Mercury

Mercury

Mercury is the closest planet to the Sun and orbits the Sun in the shortest time. Mercury rotates three times on its axis for every two revolutions around the Sun. This results in extremely hot temperatures on one side of the planet and extremely cold temperatures on the other side. Mercury has no moons.

Venus

Venus

Venus is the hottest planet. Its dense cloud cover holds in the Sun's heat and the heat given off by its volcanoes. Temperatures on Venus reach 482°C (900°F) and surface pressures are high enough to crush spacecraft. Venus also rotates backwards, and a day on Venus is longer than its year. Venus has no moons.

Earth

Earth

Earth is the water planet. It is our home. It has the right temperatures and resources for life as we know it to exist. Earth has one Moon.

Mars

Mars

The largest volcano in our solar system, Olympus Mons, is found on Mars. Mars has a thin atmosphere, but has strong winds and pink dust storms. From the surface of Mars, its two moons, Phobos and Deimos seem to move in opposite directions. Swift Phobos rises in the west and sets in the east usually twice a Martian day.

Answer these questions about the illustrations and captions above.

1. Which is the largest of the four planets?

____

2. Which planet orbits the Sun in the shortest time?

____

3. What conditions allow for life to exist on Earth?

____

____

____

4. Which planet has the largest volcano in our solar system?

____

5. Which planets have at least one moon?

____

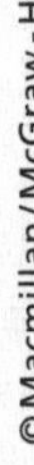

Name____________________________ Date____________

Interpret Illustrations

Lesson 2

How Do the Outer Planets Compare?

The illustrations below show three of the five outer planets. Read the captions to find out more about each planet.

Jupiter

Jupiter
The largest planet is the fastest spinner. A "day" on Jupiter is less than 10 hours long. Jupiter has a giant red spot—a storm—that has lasted over 300 years. Lightning bolts and auroras can be seen on Jupiter's night side.

Saturn

Saturn
Find an ocean big enough and this "Lord of the Rings" would float! Giant Saturn is less dense than water! Saturn has the most visible and beautiful rings of all the planets. Saturn is almost twice as far from the Sun as Jupiter is.

Uranus

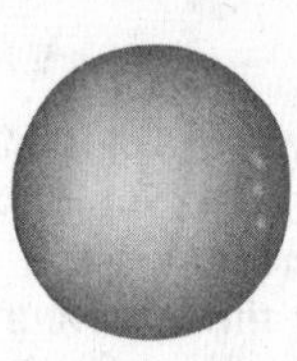

Uranus
Uranus has been called "the planet that was knocked on its side." As a result of its tilt, its poles take turns pointing toward the Sun. Even so, Uranus is hotter at its equator, though scientists don't yet know why. Uranus is about twice as far from the Sun as Saturn. A day on Uranus is $17\frac{1}{4}$ hours long, but it takes 84 Earth-years to orbit the Sun.

Answer these questions about the illustrations and captions above.

1. Which planet is farthest from the Sun?

____

2. What two occurrences can be seen on Jupiter's night side?

____

3. Why could Saturn float on water?

____

____

4. Which is longer, a day on Earth or a day on Uranus? Explain.

____

____

Name________________________ Date____________

Lesson Vocabulary
Lesson 2

The Solar System

Match the correct letter with the description.

Vocabulary
a. inner planet
b. outer planet
c. asteroid belt
d. asteroid
e. meteor
f. meteorite
g. comet
h. galaxy

_______ 1. space rocks that reach the ground

_______ 2. minor planets that orbit the Sun

_______ 3. a region between the orbits of Mars and Jupiter

_______ 4. a large group of stars held together by gravity

_______ 5. space rock that burns up before reaching the ground

_______ 6. group of planets closest to the Sun

_______ 7. mixtures of ice, rock, and dust

_______ 8. group of planets beyond the asteroid belt

Fill in the blanks to describe how the outer planets compare.

9. The ____________ lie beyond the ____________.

10. Jupiter, Saturn, Uranus, and ____________ are known as the ____________.

11. ____________ is a tiny, ____________ planet.

12. The period of ____________ is much ____________ among the outer planets.

13. The outer planets take many ____________ to orbit the ____________.

14. However, the gas giants ____________ in a period of ____________, not days.

Name ______________________ Date ______________

Cloze Test

Lesson 2

The Solar System

Vocabulary

discover	Earth	galaxy	atmospheres	NASA	giant

Fill in the blanks.

There are over billions of stars in the Milky Way ______________. Scientists are also trying to ______________ if there are other Earth-like planets. They discovered that there are more than 100 ______________ planets orbiting other stars. These giant planets are more like Jupiter than ______________. In the future, ______________ plans to launch several missions to find other planets. A possible future mission, Life Finder, would search such planets for seasonal changes in their ______________ which would indicate life on those planets.

Name______________________________ Date__________

Chapter Vocabulary

Chapter 9

Astronomy

Circle the letter of the best answer.

1. The Sun and the objects traveling around it is called the(n)
   a. planets. **b.** solar system.
   c. orbit. **d.** moons.

2. The nine objects that travel around the Sun and reflect its light are
   a. meteorites. **b.** galaxy.
   c. planets. **d.** meteors.

3. An invisible force holding the Sun and a planet together
   a. gravity. **b.** matter.
   c. distance. **d.** mass.

4. The tendency of a moving object to keep moving in a straight line is called
   a. direction. **b.** inertia.
   c. rotation. **d.** tilt.

5. One complete trip around the Sun
   a. axis. **b.** spin.
   c. rotation. **d.** revolution.

6. Stars that form patterns in the sky are called
   a. groups. **b.** pictures.
   c. light. **d.** constellations.

7. Planets that are closest to the Sun
   a. asteroids. **b.** inner planets.
   c. outer planets. **d.** debris.

Name______________________________ Date____________

8. The largest planet in our solar system
   a. Jupiter. b. Saturn.
   c. Neptune. d. Earth.

9. Planets that are beyond the asteroid belt
   a. rocky planets. b. inner planets.
   c. outer planets. d. dense planets.

10. The region between the orbits of Mars and Jupiter is the
   a. asteroid belt. b. Kuiper belt.
   c. meteor belt. d. comet belt.

11. Space rocks that burn up before reaching the ground
   a. meteorites. b. asteroids.
   c. meteors. d. comets.

12. Space rocks that reach the ground
   a. comets. b. meteorites.
   c. ice. d. meteors.

13. A mixture of ice, rock, and dust is called a(n)
   a. meteorite. b. shooting star.
   c. meteor. d. comet.

14. The Milky Way is an example of a(n)
   a. universe. b. constellation.
   c. galaxy. d. solar system.

Chapter Summary

1. What is the name of the chapter you just finished reading?

2. What are four vocabulary words you learned in the chapter? Write a definition for each.

3. What are two main ideas that you learned in this chapter?

Name ______________________ Date ____________

Chapter Graphic Organizer
Chapter 10

Weather

Find the main idea of this activity about weather. Fill in the circles and the blanks with the correct processes and supporting details. Then describe the main idea in the space provided.

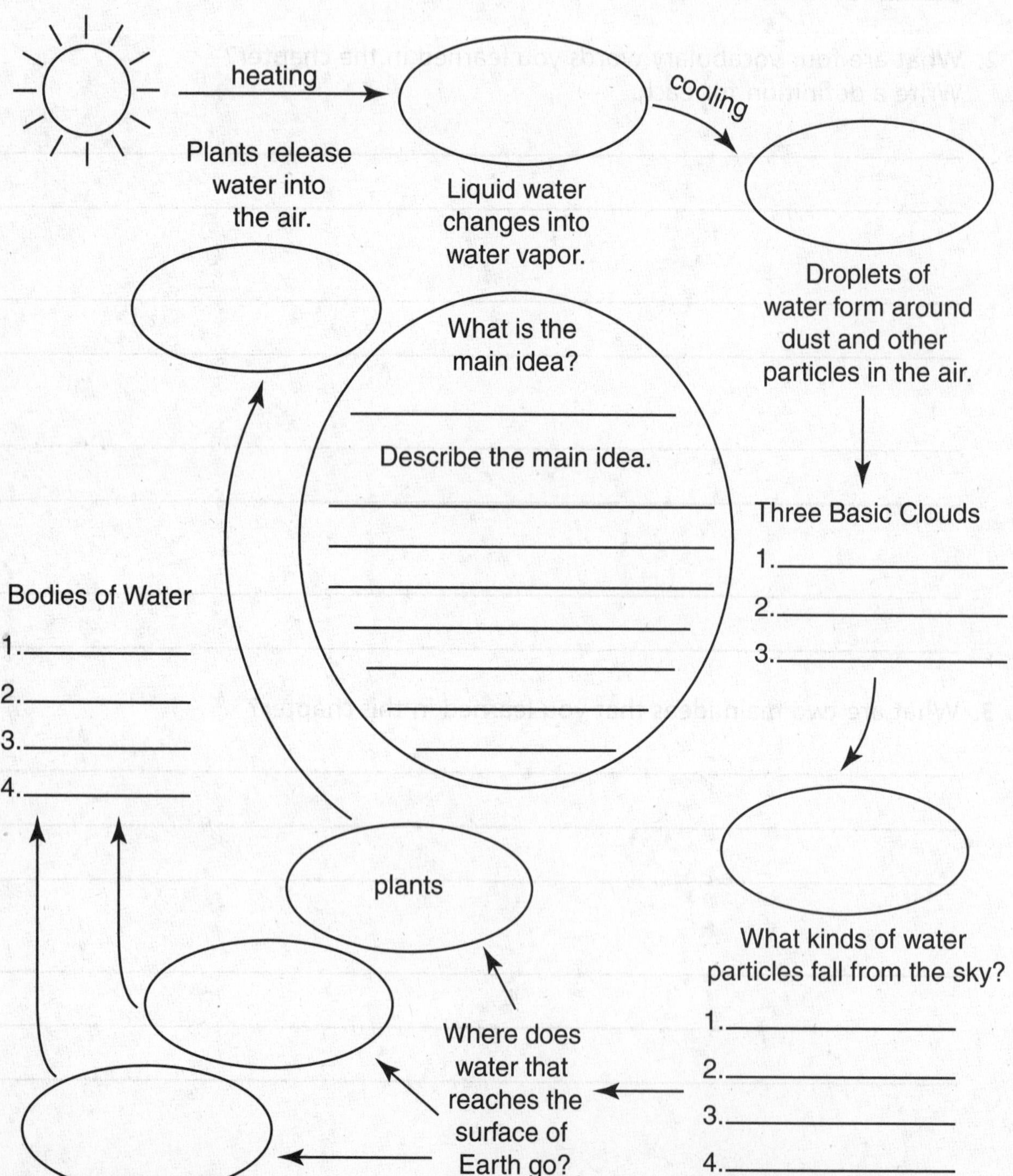

heating
cooling
Plants release water into the air.
Liquid water changes into water vapor.
Droplets of water form around dust and other particles in the air.
What is the main idea?
Describe the main idea.
Three Basic Clouds
1.
2.
3.
Bodies of Water
1.
2.
3.
4.
plants
What kinds of water particles fall from the sky?
1.
2.
3.
4.
Where does water that reaches the surface of Earth go?

Main Idea

Every story has a main idea, a sentence that tells what the story's mostly about. Other facts in the story are details that support, or tell more about, the main idea. Read the following paragraphs, based on information from this chapter of your textbook. Then circle the main idea and underline the supporting details in each paragraph.

1. What happens to the Sun as time goes by each day? In the morning the Sun is close to the horizon. By noon the Sun is high in the sky above you, as high as it gets all day. After noon the Sun begins to get lower and lower in the sky until it seems to disappear below the horizon.
2. Air temperature drops with altitude at about 2°C (3.5°F) every 305 meters (1,000 feet). For example, in Lewiston, Maine, elevation 34 meters (110 feet), it's a pleasant 21°C (70°F). Two hours away in Mount Washington, New Hampshire, elevation 1,917 meters (6,288 feet), it's only 1°C (34°F).

Now reread the Science Magazine article "Flood: Good News or Bad?" Then write the main idea of the article, and list at least four supporting details.

Main Idea: ______________________________

____

Supporting Details:

1. ______________________________

____

2. ______________________________

____

3. ______________________________

____

4. ______________________________

____

Name ______________________ Date ____________

Chapter Reading Skill

Chapter 10

More Ideas and Details

Read the following paragraph about the monsoons. Then circle the main idea, and underline each of the supporting details.

> Why do the monsoons change direction? In summer the Sun heats dry air over tropical land, while nearby oceans stay cooler. The warm air rises above the land, and cooler air over the ocean blows in to take its place. The ocean wind brings heavy rain to the land. An area may get as much as 2.54 meters (100 inches) of rain a month! Then in winter the Sun heats the land less, so it cools off. As warm air rises over the ocean, dry cool wind from the land blows out to take its place. That same land area may get only 2.54 centimeters (1 inch) a month!

Now write a main idea and some supporting details about clouds, air pressure, the atmosphere, the water cycle, relative humidity, sea and land breezes, or some other weather condition. Give your paper to a classmate, and have him or her use the information to write a paragraph.

Main Idea: ______________________

____

Supporting Details:

____

____

____

____

____

Name ____________________ Date __________

Lesson Outline

Lesson 3

Atmosphere and Air Temperature

Fill in the blanks.  Reading Skill: **Main Idea** - questions 1, 2, 16

Does the Sun's Angle Matter?

1. The hottest parts of Earth are the areas near ____________, while the coldest are the areas ____________.
2. The angle of the Sun's rays is the cause of differences in temperatures at the ____________ and the ____________.
3. Combining parts of the words "incoming," "solar," "radiation," forms the term "____________."
4. The greater the angle of insolation, the ____________ a place will be.

What Affects Insolation?

5. The Sun is highest in the sky at ____________.
6. The angle of insolation can be measured by examining the angles created by ____________.

Why Do You Cool Down as You Go Up?

7. Air temperatures drop as you go higher up a mountain because you are traveling higher above ____________.
8. Temperatures drop about ____________ degrees for every ____________ in altitude.
9. The troposphere is the lowest layer of Earth's ____________.

Name ______________________ Date __________

Lesson Outline

Lesson 3

What Happens to the Air Pressure?

10. As you go higher in altitude, ______________ decreases steadily.

11. Air is made up mostly of ______________ of the gases ______________ and ______________.

12. Normal air pressure is greatest at ______________, because the ______________ is tallest there.

13. Besides gases, air contains ______________ and ______________.

What Is Weather?

14. The term "weather" really describes what conditions are like in the ______________________ at a given place and time.

15. Besides temperature and air pressure, weather characteristics that change are

    a. ______________________,

    b. ______________________,

    c. ______________________, and

    d. ______________________.

16. Thermometers measure temperature by either the ______________ or the ______________ scale.

17. In a mercury barometer, ______________ pushes mercury up in a tube.

18. A spring inside an accordion-like metal can measures air pressure in a(n) ______________ barometer.

Name______________________ Date__________

Does the Sun's Angle Matter?

Diagrams like this one take careful study. Notice the direction of the arrows, which show how Earth moves around the Sun during a year. Then notice that the diagram shows Earth's position and tilt relative to the Sun at different times of the year. To understand the diagram, look carefully at the changes from season to season. Observe the differences in the angle at which the Sun's rays fall on different parts of Earth.

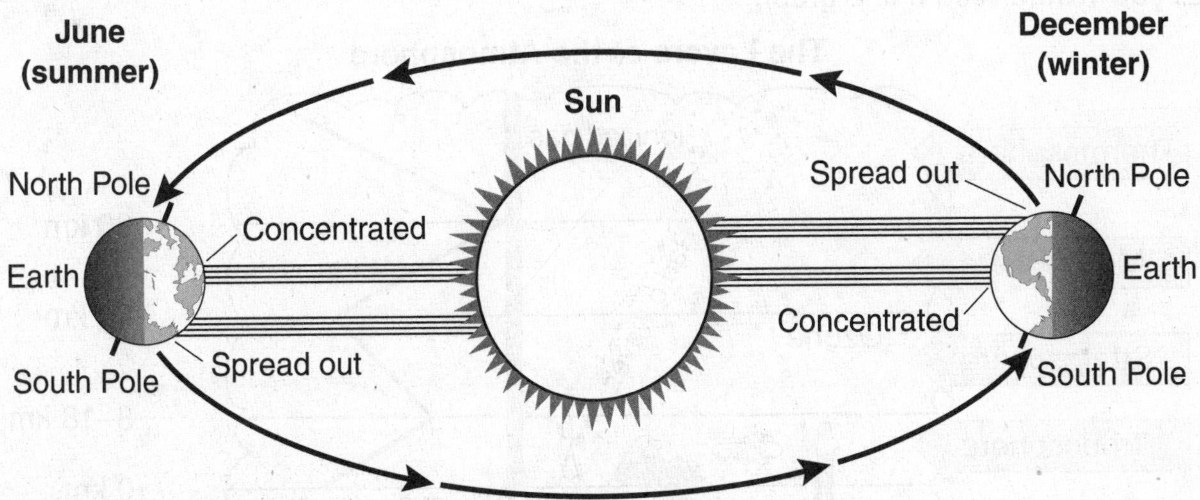

Answer these questions about the diagram above.

1. In December, which half of Earth, northern or southern, is tilted away from the Sun?

   ____

   Are the Sun's rays more or less concentrated in that half in December?

   ____

2. In June, is the southern half of Earth tilted toward or away from the Sun?

   ____

3. What part of Earth receives concentrated sunlight all year round?

   ____

4. Where on Earth are the Sun's rays spread out thinly all the time?

   ____

Name ______________________ Date ____________

Interpret Illustrations

Lesson 3

Why Do You Cool Down as You Go Up?

The two parts of this diagram show different, but related, kinds of information. To understand it, you need to look carefully at how the two parts are related. The left-hand part shows the layers of the atmosphere and their heights above the Earth. The other part is a graph of temperature change in each layer. Read it as you would read a line graph.

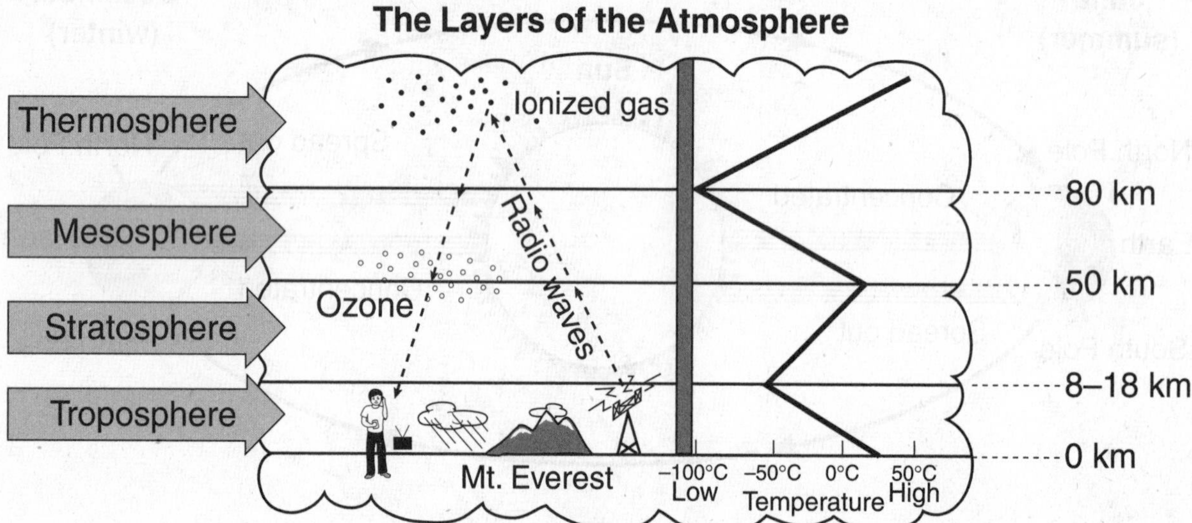

Answer these questions about the diagram above.

1. In what layer of the atmosphere do people live and clouds form?

   ____

   At about what height is the top of that layer? ____________

2. Describe the location of the ozone layer.

   ____

3. Do temperatures get warmer or colder as you move from the lower stratosphere to the top of that layer? ____________

   How big is the change? ______________________

4. The topmost layer of the atmosphere is the ____________.

   How high above Earth does this layer begin? ____________

5. Which layer has the greatest temperature change? ____________

Name ______________________ Date ____________

Lesson Vocabulary
Lesson 3

Atmosphere and Air Temperature

Vocabulary

- insolation
- weather
- nitrogen
- air pressure
- troposphere
- barometer
- oxygen
- atmosphere
- temperature

Fill in the blanks.

1. The air that surrounds the Earth is the ____________.
2. Ninety-nine percent of the air we breathe is made up of ____________ and ____________.
3. Air pressure is measured by a(n) ____________.
4. The levels of the four layers of the atmosphere are determined by sudden changes in ____________ at each level.
5. The amount of the Sun's energy that reaches Earth at a given place and time is ____________.
6. All the moisture around Earth is in the ____________.
7. The force put on a given area by the weight of the air above it is ____________.
8. The conditions present in the troposphere at any given time determine the ____________.

Answer each question.

9. How does air pressure change with altitude? Why?

____

____

10. What happens to temperature as the angle at which the Sun's rays strike Earth's surface increases? As the angle decreases? What happens to the length of shadows at the same time?

____

____

____

Name ____________________ Date ____________

Cloze Test
Lesson 3

Atmosphere and Air Temperature

Vocabulary

thermometer	Celsius	air pressure
weight	aneroid	air
spring	changes	barometer

Fill in the blanks.

You can measure temperature using a ____________________. The temperature is read on a Fahrenheit or a(n) ____________________ scale. ____________________ can be measured using a mercury ____________________, which has a glass tube with the open end submerged in liquid mercury. The mercury stops rising when its ____________________ equals the air pressure. You also can use a(n) ____________________ barometer to measure air pressure. This device is an accordion-like metal can with most of the ____________________ removed. Inside, a(n) ____________________ balances the outside air pressure. A needle indicates the ____________________ in air pressure.

Name_______________________ Date___________

Water Vapor and Humidity

Fill in the blanks.  Reading Skill: **Main Idea** - questions 2, 5, 11

Where Does Water Vapor Come From?

1. When water drops form on a cold glass, the water comes from the ____________________.
2. Water vapor is water in the form of a(n) ____________________.
3. The amount of water vapor in the air is ____________________.
4. Water covers more than ____________________ of Earth's surface.
5. Liquid water becomes a gas called water vapor through the process of ____________________.
6. The Sun's energy enables ____________________ from the water's surface to escape into the atmosphere.
7. The change from a gas to a liquid is called ____________________.
8. The second-largest source of water vapor is ____________________ from plant leaves.

Name______________________________ Date______________

Lesson Outline

Lesson 4

What Happens When Warm, Moist Air Cools?

9. Whenever warm air rises higher in the sky, it ______________.

10. When cool and warm air meet, the lighter ______________ is pushed up over the heavier, cold air.

11. Whenever warm air rises and cools, the water vapor in the air ______________.

12. Water droplets in the air form ______________.

Name ______________________ Date ____________

Interpret Illustrations

Lesson 4

Where Does Water Vapor Come From?

A map is a kind of diagram. This map shows Earth's water and land. Notice how the areas of water compare with the areas of land.

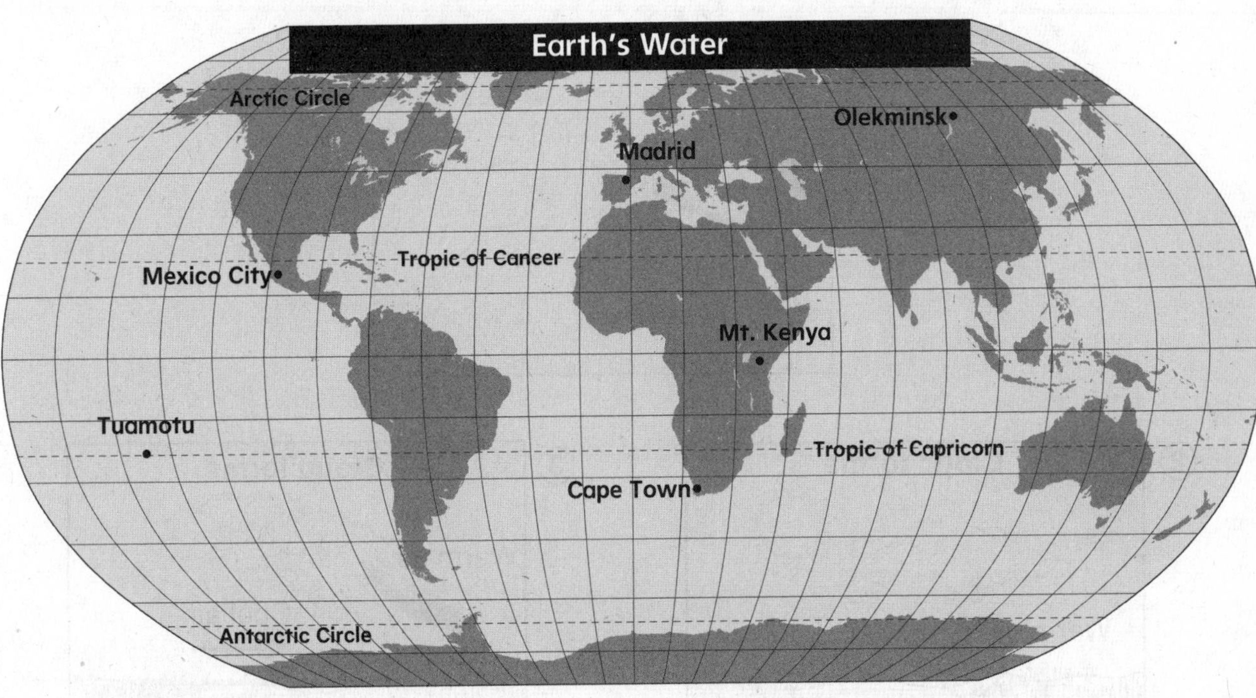

Answer these questions about the diagram above.

1. What is the title of the diagram?

   ____

2. What do the lighter gray areas represent?

   ____

3. What do the darker gray areas represent?

   ____

4. More than two-thirds of Earth is covered with water. Do you think the diagram shows this accurately? Explain.

   ____

   ____

How Clouds Form

These three diagrams show three ways in which clouds are formed when water vapor in the air condenses. When air moves higher, it cools. Because the temperature decreases, condensation from water droplets is greater than evaporation from the droplet. As a result, the droplets grow. The diagrams show what can happen next.

① 

② 

③ 

Answer these questions about the diagram above.

1. In diagram 1, does the air get warmer or cooler when the wind pushes it up over the mountain? ____________________

2. What happens to the water vapor in the air when the air cools?

____

3. In diagram 2, the Sun warms the ground. What happens to the warm air near the ground?

____

4. In diagram 3, the warm air is pushed up by the ____________ below it.

Name ______________________ Date ____________

Lesson Vocabulary
Lesson 4

Water Vapor and Humidity

Vocabulary
humidity
evaporation
relative humidity
condensation
transpiration
water vapor

Fill in the blanks.

1. The amount of water vapor in the air is called the ____________________.
2. The changing of a liquid into a gas is called ____________.
3. Water in the form of a gas is ____________________.
4. The changing of a gas into a liquid is ____________________.
5. A comparison between how much water vapor is in the air and how much the air could hold is called ____________________.
6. The leaves of plants release water vapor into the air through ____________________.

Answer each question.

7. What is relative humidity?

____

____

8. What two factors determine the amount of humidity in the air?

____

____

____

9. What effect does temperature have on evaporation?

____

____

10. When does condensation occur?

____

____

Name ______________________ Date __________

Cloze Test

Lesson 4

Water Vapor and Humidity

Vocabulary

humidity	**evaporate**	**water vapor**	**gas**	**two-thirds**
oceans	**trillions**	**plants**	**absorb**	**evaporation**

Fill in the blanks.

The water in the air is ______________. Water vapor is water in the form of ______________. ______________ is the amount of water vapor in the air. More than ______________ of planet Earth is covered with liquid water from ______________, rivers, and lakes. There is also water in the ground, and in ______________. The changing of liquid water into a gas is called ______________. Each day the Sun turns ______________ of tons of ocean water into water vapor. Water molecules ______________ the Sun's energy and speed up. "Speedy" water molecules near the surface of the liquid "escape" or ______________ in the atmosphere as water vapor.

Name ______________________ Date __________

Lesson Outline

Lesson 5

Clouds and Precipitation

Fill in the blanks.  Reading Skill: **Main Idea** - questions 2, 5

How Do Clouds Form?

1. Clouds are made up of ______________ or ______________.
2. All clouds form when the air cools and water vapor ______________ around dust and other particles.
3. A thick, sharp-edged gray cloud is probably made of ______________, not ______________.
4. All clouds form in the lower atmosphere, or ______________.
5. The three basic cloud forms are:
   a. blanketlike layers, or ______________,
   b. puffy, flat-based ______________, and
   c. high-altitude, feathery ______________.
6. The prefix "nimbo-," or nimbus, as in nimbostratus, refers to a cloud that brings ______________ or ______________.
7. A cloud at ground level is called ______________.

What Is Precipitation?

8. Any form of water particles that falls to the ground from the atmosphere is ______________.
9. Precipitation occurs when water droplets or ice crystals join together and become ______________.
10. A dust particle is the ______________ around which water molecules condense.
11. The type of precipitation formed from ice crystals that occurs when the ground temperature is cold is ______________.
12. When cloud droplets collect and freeze around an ice crystal, and the process repeats over and over, ______________ is formed.

Name ____________________ Date ____________

Lesson Outline

Lesson 5

Fill in the blanks.

Are Cloud Type and Precipitation Related?

13. The shape and height of certain clouds determine the kind of ________________ that falls.

14. Rain or snow from large ________________ clouds is likely to be heavier but usually doesn't last as long as that from ________________ clouds.

15. Heavy rain and sometimes ________________ come from clouds whose tops reach where it is below freezing.

16. Strong winds, or ________________, inside a huge cloud move ice crystals upward again and again.

How Do You Record How Cloudy It Is?

17. A term like "overcast" refers to ________________, the portion of the sky covered by clouds.

18. In a weather station model, different amounts of cloud cover is shown by ________________________.

Name ______________________ Date __________

Interpret Illustrations

Lesson 5

How Do Clouds Form?

Some diagrams show parts of a whole. The diagram below shows the basic types of clouds found in the atmosphere. Notice how the cloud forms are related to each other.

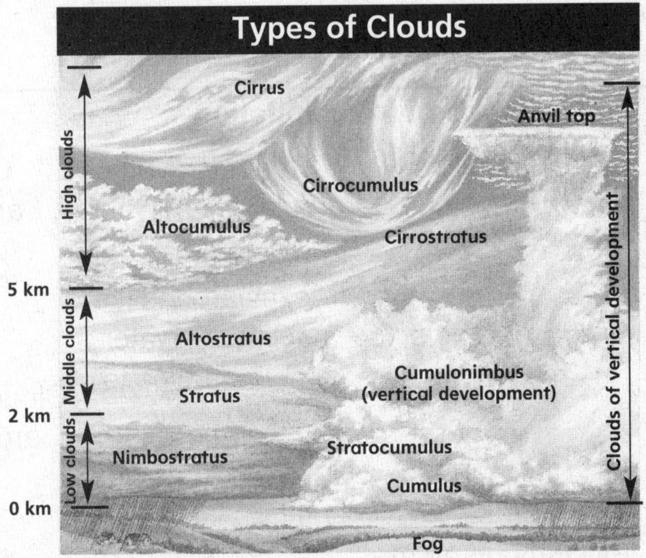

Answer these questions about the diagram above.

1. What are the four general families of clouds shown in the diagram?

____

____

2. What is the upper height range for low clouds?

____

3. Which two cloud forms are found in the middle cloud family?

____

4. What are ground-level clouds called?

____

5. How are clouds of vertical development different from the other three families of clouds?

____

____

Name ____________________ Date __________

Interpret Illustrations

Lesson 5

What Is Precipitation?

Charts like this one organize information. This chart shows four different types of precipitation and how each one forms. To understand the chart, read the labels on each column. Then follow the arrows to see the steps in forming each kind of precipitation.

Types of Precipitation

Rain	Sleet	Snow	Hail
Condensation around nucleus	Condensation around nucleus	Condensation around ice nucleus	Condensation around nucleus
Cloud droplets collect.	Cloud droplets collect.	Supercooled water freezes around ice nucleus or water vapor changes to ice crystals.	Cloud droplet
Fall through warm air	Raindrop	Ice crystals grow larger.	Freezing
	Fall through air at freezing temperature	Snowflakes	Cloud droplets collect around ice crystal
		Cold ground temperature	Freezing
			Repeats over and over.
			Warm ground temperature

Answer these questions about the diagram above.

1. What first step do all types of precipitation have in common?

____

2. When raindrops fall through air at freezing temperatures, they become

____

3. What two things can happen after condensation forms around an ice nucleus?

____

____

4. A snowflake forms when ______________ grow larger.

Name ____________________ Date __________

Lesson Vocabulary

Lesson 5

Clouds and Precipitation

Vocabulary

stratus
fog
cumulus
cirrus
temperature
precipitation

Fill in the blanks.

1. Moist air cooled at ground level causes ____________ to form.
2. High-altitude clouds that look feather-like are ____________ clouds.
3. Whether precipitation is rain or snow depends on the ____________ in the atmosphere and on the ground.
4. A cloud that looks puffy but has a flat bottom is a(n) ____________ cloud.
5. Rain, snow, sleet, and hail are all kinds of ____________.
6. Clouds that form in blanketlike layers are ____________ clouds.

Answer each question.

7. What kind of precipitation would you expect from stratus clouds?

____

____

8. What are the five terms used to describe how cloudy it is on a given day?

____

____

Name ______________________ Date ____________

Cloze Test
Lesson 5

Clouds and Precipitation

Vocabulary

cirrus	stratus	fog
height	nimbo	cumulonimbus
cumulus	troposphere	shape

Fill in the blanks.

Clouds are found only in the ________________. Those that form in blanketlike layers are ________________ clouds. ________________ clouds form at very high altitudes out of ice crystals. They have a featherlike ________________. Puffy clouds that seem to rise up from a flat bottom are called ________________ clouds. If snow or rain falls from a cloud, the term ________________ is added to the cloud's name. Clouds are further grouped into families by form and ________________. For example, ________________ clouds develop upward. A cloud at ground level is called ________________.

Name________________________ Date____________

Lesson Outline
Lesson 6

Air Pressure and Wind

Fill in the blanks.  Reading Skill: **Main Idea** - questions 3, 13, 14

How Can Air Pressure Change?

1. If the volume of a container increases, the air pressure inside the container will ________________.
2. Air pressure is ________________ at sea level, than high in the atmosphere.
3. When air is heated, air pressure decreases because the molecules ________________ and ________________.
4. The same volume of air ________________ less, and the pressure ________________.
5. Air contains the gases ________________ and ________________, which are heavier than molecules of water vapor. So, moist air exerts ________________ pressure than dry air.
6. In a weather station model, a line slanting up and to the right shows that air pressure is ________________.

Why Do Winds Blow?

7. When denser air moves toward less dense air, it creates ________________.
8. Warm air rises because it is ________________, or less dense.
9. Rising air is a(n) ________________, while sinking air is a(n) ________________.
10. Rising warm air and sinking cool air form a circular wind pattern called a(n) ________________.
11. Convection cells form because of unequal ________________ and ________________ of the air.

Name ________________________ Date ____________

Lesson Outline

Lesson 6

Fill in the blanks.

What Are Sea and Land Breezes?

12. Land warms up ______________ than water.

13. On a warm day, the air above the land ______________, and a cool ______________ blows toward the land.

14. At night, the land cools more quickly than the water, causing a(n) ______________ to blow toward the water.

15. A valley breeze blows up a mountain slope because the air on the slope is ______________.

What Is the Coriolis Effect?

16. Winds in the Northern Hemisphere seem to curve to the ______________ because Earth is ______________ under them.

How Are Global Wind Patterns Produced?

17. Warm, moist air over the equator creates a zone of ______________, while cold dry air at the poles has ______________.

18. Global wind zones are set up by air moving from zones of ______________ to zones of ______________.

19. Global winds seem to travel in curved paths because of the ______________.

What Are Isobars?

20. A line on a weather map connecting locations with equal air pressure is called a(n) ______________.

Name______________________________ Date____________

Interpret Illustrations

Lesson 6

How Can Air Pressure Change?

This diagram illustrates an experiment showing that volume is one factor that causes changes in air pressure. It shows two jars with a plastic bag fastened over their tops. Look at the jars and notice the difference between the air molecules inside them. Then study the smaller diagrams, which show the pressures inside and outside the jars.

An Air Pressure Model

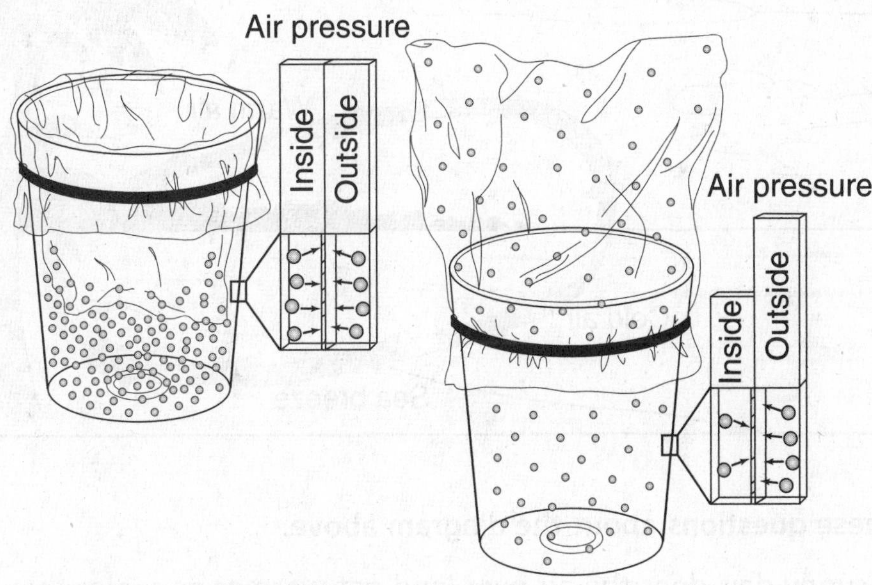

Answer these questions about the diagram above.

1. In the jar on the left, is the inside air pressure greater than, less than, or the same as the outside air pressure? ____________________

2. Pulling up the plastic bag on the right-hand jar increases the ____________________ inside the bag-jar system.

3. Are the air molecules in the right-hand jar closer together or farther apart than those in the other jar? ____________________

4. When the volume increases, the air pressure in the jar becomes ____________________ than the air pressure outside.

Name________________________ Date______________

Interpret Illustrations

Lesson 6

What Are Sea and Land Breezes?

Land temperatures change faster than water temperatures do. This diagram shows how those temperature changes can affect winds and wind directions along a coastline. This diagram shows how a sea breeze—a breeze that blows from the ocean—is created during the day.

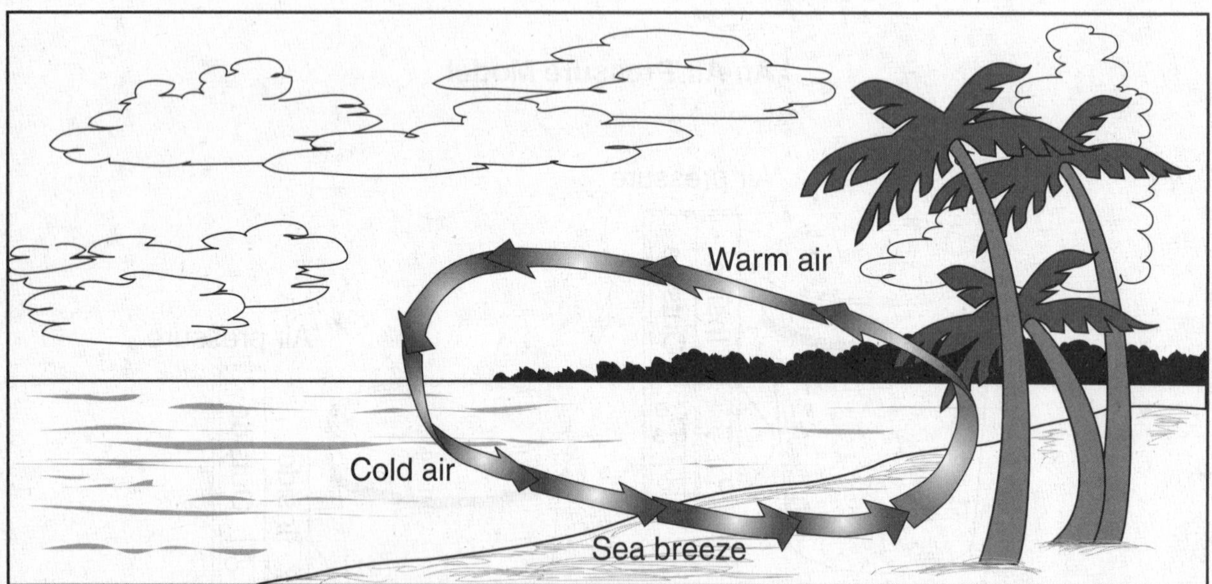

Answer these questions about the diagram above.

1. On a sunny day, does the air over land get warmer or cooler than the air over the water? ______________
2. In what direction does the warmer air move? ______________
3. When a sea breeze blows, it brings ______________ air to the land.
4. This diagram shows why a sea breeze blows inland during the day. Now figure out what will happen to these wind patterns at night. Remember that land temperatures change faster than water temperatures. Explain what you think will happen.

____

____

Name______________________ Date__________

Lesson Vocabulary
Lesson 6

Air Pressure and Wind

Vocabulary
convection cell
Coriolis effect
wind
isobar
land breeze
sea breeze

Fill in the blanks.

1. Air that moves horizontally is called ______________.
2. A breeze that blows from sea to land is a(n) ______________.
3. A line on a map connecting places with equal air pressure is a(n) ______________.
4. A breeze that blows from land to sea is a(n) ______________.
5. In the Northern hemisphere the wind curves to the right; this curving is called the ______________.
6. The pattern of air rising, sinking, and blowing horizontally is a(n) ______________.

Answer each question.

7. What are some of the factors that might interrupt the wind blowing in a straight line forever?

____

____

8. Compare the air pressure at sea level with the air pressure on top of a mountain. Why are the two pressures different?

____

____

____

Name ______________________ Date ____________

Cloze Test

Lesson 6

Air Pressure and Wind

Vocabulary

winds	counterclockwise	shorter	southern
Coriolis effect	northern	poles	rotation

Fill in the blanks.

Earth's ______________ affects ______________ blowing across its surface. Places near the ______________ travel a ______________ distance than places near the equator. Earth rotates ______________ as seen from the North Pole. In the ______________ Hemisphere, winds curve to the right. In the ______________ Hemisphere, winds curve to the left. The ______________ is the name given for these curving effects.

Name________________________________ Date____________

Chapter Vocabulary

Chapter 10

Weather

Circle the letter of the best answer.

1. As you get higher in altitude, air pressure

a. increases. **b.** remains the same.

c. decreases. **d.** rises.

2. The layer of atmosphere closest to Earth's surface is the

a. mesosphere. **b.** stratosphere.

c. troposphere. **d.** ozone.

3. When weather is described as overcast

a. there are no clouds.

b. the entire sky is covered by clouds.

c. there are a few clouds in the sky.

d. the sky is half covered by clouds.

4. Insolation is

a. the amount of the Sun's rays that reach Earth.

b. the angle at which the Sun's rays hit Earth.

c. the insulating effect of the atmosphere.

d. another name for the mesosphere.

5. Humidity is

a. found mostly along shorelines.

b. how much moisture the air could hold if it was warm.

c. the amount of water vapor in the air.

d. the amount of water vapor at ground level.

6. An instrument used to measure air pressure is a(n)

a. barometer.

b. thermometer.

c. gauge.

d. balance.

7. A circular pattern of updrafts, downdrafts, and winds is a(n)
   - a. weather pattern.
   - b. convection cell.
   - c. Coriolis effect.
   - d. isobar.

8. Wind is air that moves
   - a. vertically.
   - b. in a circular pattern.
   - c. horizontally.
   - d. down.

9. Cumulonimbus clouds are
   - a. pretty, fluffy clouds.
   - b. clouds that bring thunderstorms.
   - c. whispy clouds made of ice crystals.
   - d. clouds at ground level.

10. Which is NOT a form of precipitation?
   - a. rain
   - b. snow
   - c. hail
   - d. lava

11. A wind that blows from the ocean onto land is called a(n)
   - a. sea breeze.
   - b. mountain breeze.
   - c. land breeze.
   - d. valley breeze.

Chapter Summary

1. What is the name of the chapter you just finished reading?

____

____

2. What are four vocabulary words you learned in the chapter? Write a definition for each.

____

____

____

____

____

____

____

____

3. What are two main ideas that you learned in this chapter?

____

____

____

____

____

____

Name________________________________ Date____________

Chapter Graphic Organizer
Chapter 11

Weather Patterns and Climate

Listed below are 12 events that occur during a thunderstorm. Put the events in the proper order from beginning to end. Starting with the beginning of a thunderstorm, write the number of each event in the correct circle.

1. Heavy rain turns to light rain and then stops.
2. An upward rush of air called an updraft causes clouds to form.
3. Falling rain causes air to move downward in a downdraft.
4. A large buildup of static electricity releases a huge spark called lightning.
5. Air expands with such force that it creates a very loud sound called thunder.
6. Heavy rain or hail falls to the ground.
7. Intense heating causes warm, moist air to rise quickly.
8. Updrafts can no longer support water droplets or ice crystals.
9. Downdrafts become stronger than updrafts.
10. Falling air rubbing against rising air causes static electricity to build up.
11. Updrafts continue, causing clouds and water droplets to grow larger and larger.
12. Lightning superheats the air, causing it to expand suddenly.

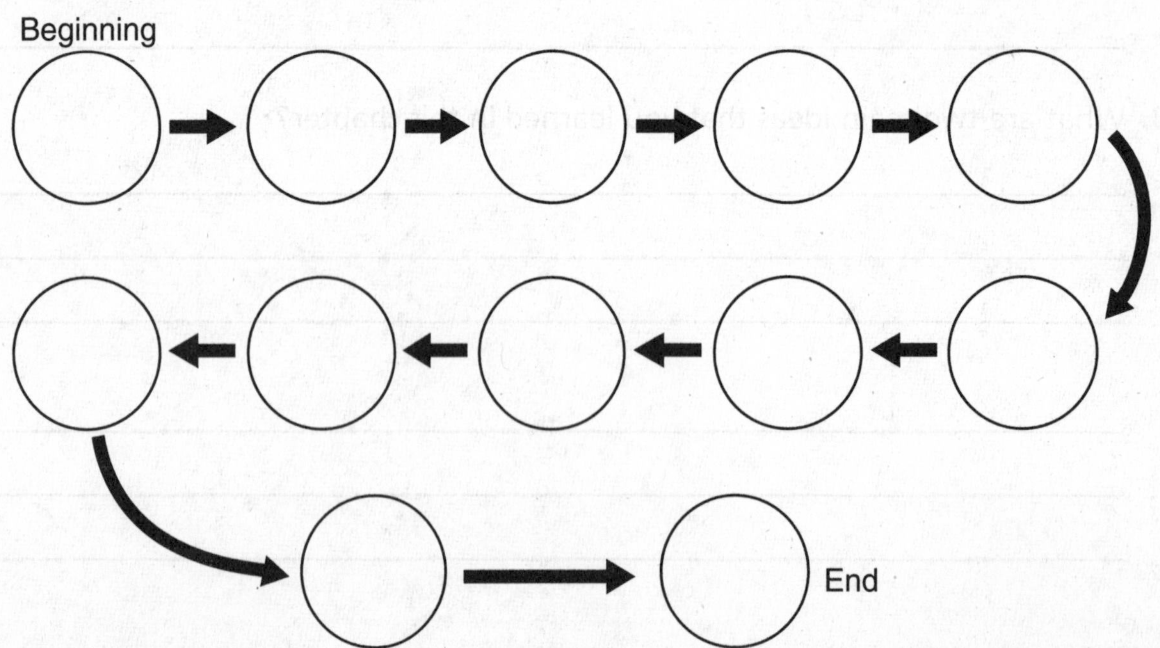

Name ______________________ Date __________

Chapter Reading Skill

Chapter 11

Sequence of Events

Everything in life happens in sequence. You can't have your 13th birthday until you've had your 12th birthday, you can't eat pizza until it comes out of the oven, and you can't wake up until you go to sleep!

Read each example below. Then answer the questions.

1. The money we raised at the bake sale paid for our trip to Washington.

   What happened first? ______________________

   What happened second? ______________________

2. We had a victory celebration after our team won the championship game.

   What happened first? ______________________

   What happened second? ______________________

3. I had to baby-sit my cousin in the morning, but later I baked cookies and took them to Pat's party.

   What happened first? ______________________

   What happened next? ______________________

   What happened last? ______________________

4. I read the book, then researched the author on the Internet, and finally finished my report at 10 P.M.!

   What happened first? ______________________

   What happened next? ______________________

   What happened last? ______________________

5. We set up our tents and went swimming. After lunch we went hiking in the woods.

   What happened first? ______________________

   What happened next? ______________________

   What happened next? ______________________

   What happened last? ______________________

Name ______________________ Date __________

Chapter Reading Skill
Chapter 11

In Sequence

A flowchart is a good way to show how warm, moist air can travel over a mountain and affect a mountain's climate.

Use the sentences below to fill in the blank flowchart.

1. Hot, dry air descends down the leeward side of the mountain.
2. Cold, moist air on top of mountain.
3. Warm, moist air blows up the side of the mountain.
4. It begins to rain on the windward side.
5. Moist air blows in from the ocean.
6. Air loses moisture.

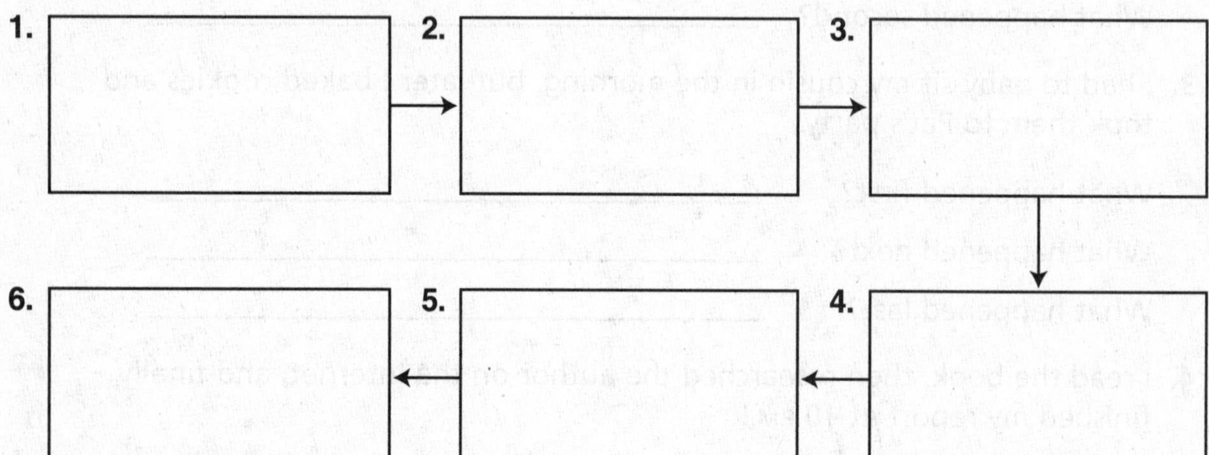

Name________________________________ Date______________

Lesson Outline
Lesson 7

Air Masses and Fronts

Fill in the blanks.  Reading Skill: **Sequence of Events** - questions 4, 11

How Do Air Masses Affect Weather?

1. A large region of the atmosphere where the air has similar properties throughout is called a(n) ____________________.
2. An air mass that forms over the warm water of the Gulf of Mexico will be ____________________ and ____________________.
3. Once an air mass forms, it is moved by ____________________.
4. In the United States, air masses generally move from ____________________ to ____________________.
5. At the boundary between air masses with different temperatures a(n) ____________________ forms.
6. A continental tropical air mass will bring ____________________ air to an area.
7. A continental polar air mass will bring ____________________ air to an area.
8. A maritime tropical air mass will bring ____________________ air to an area.
9. A maritime polar air mass will bring ____________________ air to an area.

Name ______________________________ Date ______________

Lesson Outline

Lesson 7

How Do Fronts Affect Weather?

10. Cold air moves in under a warm air mass, often bringing winds and brief, heavy storms, in a(n) ____________________.

11. After a cold front passes, the weather is usually ____________________, ____________________, and ____________________.

12. Warm air moves up and over a cold air mass in a(n) ____________________.

13. Rainy or snowy days and warmer, more humid weather is brought by ____________________.

Use with textbook pages D68–D73

Name________________________ Date____________

Interpret Illustrations

Lesson 7

How Do Air Masses Affect Weather?

This diagram shows how a weather front forms when warm and cold air masses meet. To understand a diagram like this, read all the labels. Notice the positions of the warm and cold air. Then follow the arrows to trace the movements of the different air masses.

A Weather Front

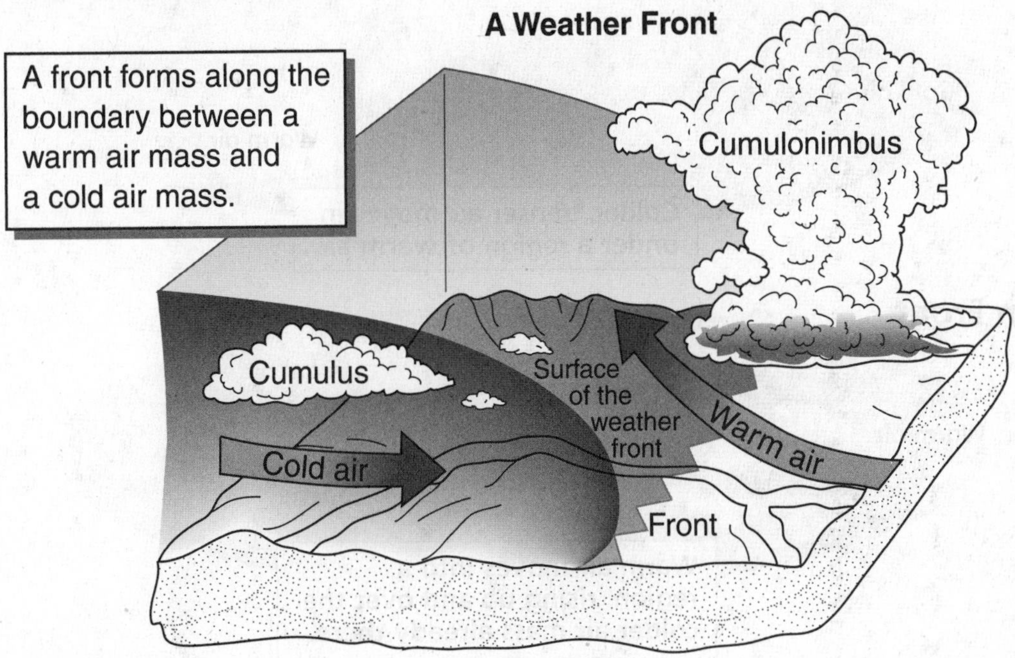

Answer these questions about the diagram above.

1. Is the cold air mass moving toward the left or the right side of the diagram?

   ____

2. How does the warm air move when it meets the cold air mass?

   ____

3. What kinds of clouds form where the warm air is rising?

   ____

4. What kind of clouds form behind the front?

   ____

5. What do the arrows along the front show?

   ____

Name ______________________ Date ____________

Interpret Illustrations

Lesson 7

How Do Fronts Affect Weather?

These diagrams show how two different kinds of fronts—a cold front and a warm front—affect the weather in an area. Read all the labels. Notice the arrows that show the direction in which air masses are moving. Look for differences between the two kinds of fronts.

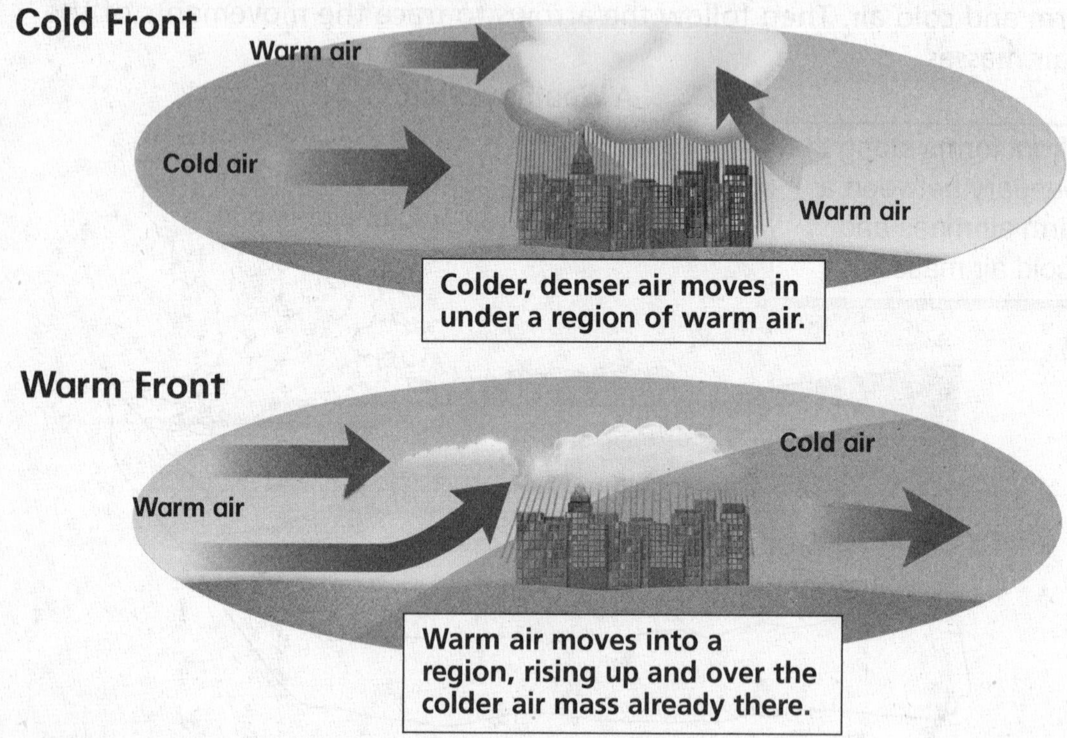
Cold Front
Warm air
Cold air
Warm air
Colder, denser air moves in under a region of warm air.
Warm Front
Cold air
Warm air
Warm air moves into a region, rising up and over the colder air mass already there.

Answer these questions about the diagram above.

1. In the diagram of a cold front, what happens to the warm air as the cold air mass moves in? ______________________

2. What happens to cold air when warm air moves into a region?

____

3. In both diagrams, what happens when warm air meets cold air?

____

4. How is the way cold air moves into a warm air region different than the way warm air moves into a cold air region?

____

____

Name ______________________ Date ____________

Lesson Vocabulary
Lesson 7

Air Masses and Fronts

Fill in the blanks.

Vocabulary
cold front
east
west
air mass
front
warm front

1. A large region of the atmosphere where the air has similar properties throughout is called ______________.
2. A narrow boundary between two air masses is called a(n) ______________.
3. Brief, heavy storms are caused by a(n) ______________.
4. Light, steady rain or snow is caused by a(n) ______________.
5. The global winds in the United States blow from ______________ to ______________.

Answer each question.

6. Why are fronts so important to our everyday life?

____

____

____

7. What do scientists use to help them predict weather?

____

____

____

Name______________________________ Date______________

Cloze Test
Lesson 7

Air Masses and Fronts

Vocabulary

air mass	cold	snow	warm
drier	cooler	thunderstorm	

Fill in the blanks.

In a(n) ________________ front, cold air moves in under a warm ________________. Cold fronts often bring brief, heavy ________________ and strong winds. After the storm the skies are usually clearer, and the weather is ________________ and________________.

In a(n) ________________ front, warm air moves in over a cold air mass. Warm fronts often bring light, steady rain or ________________.

Name ______________________ Date ____________

Lesson Outline

Lesson 8

Severe Storms

Fill in the blanks.  Reading Skill: **Sequence of Events** - questions 1, 3, 5, 8, 13

What Are Thunderstorms?

1. In a thunderstorm, ________________ move warm air upward in a cloud.
2. Thunderstorms form in ________________ clouds.
3. When there is too much water in a thunderstorm cloud, it falls as ________________.
4. Falling rain causes ________________ in a thunderstorm cloud, building up ________________.
5. A spark between electric charges in a cloud is ________________.
6. Thunderstorms are likely to form in the warm air ahead of a(n) ________________.
7. Air rushes in from all sides when the updraft in a(n) ________________ ________________ is very strong.
8. As the air curves into a spin, the tornado gets stronger, and a(n) ________________ forms.
9. In the center of a tornado, winds can reach a speed of ________________.
10. The regions of the United States where tornadoes are most likely to occur are:

    a. ________________ and

    b. ________________.

Name ______________________________ Date ____________

Lesson Outline

Lesson 8

How Do Hurricanes Form?

11. A hurricane is a large swirling storm with an area of very ________________ at the center.

12. A low-pressure center forms in a hurricane because of strong heating and lots of ________________ from the surface of warm ocean water.

13. When the wind nears the center, it moves upward and forms a(n) ________________ of tall thunderstorms.

14. Winds spiral counterclockwise in the Northern Hemisphere because of the ________________.

15. Hurricanes have a wind speed of ________________________ or higher.

16. An area of light winds and clear skies at the center is the ________________ of a hurricane.

17. The low air pressure beneath a hurricane causes a rise in the sea called a(n) ________________.

How Can Radar Track Storms?

18. The word *radar* stands for radio, ________________, and ________________.

19. Doppler radar uses ________________ to locate storms.

Name________________________________ Date______________

Interpret Illustrations
Lesson 8

What Are Thunderstorms?

These two diagrams show how a thunderstorm forms. As with any diagram, it's important to start by reading all the labels and captions. Also remember that arrows usually show movement or steps in a process. Notice how the arrows change in each diagram.

① Strong updrafts form inside the cloud.

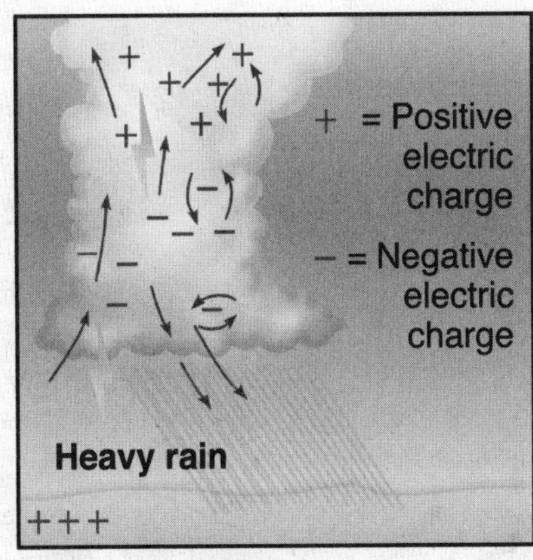

② Electric charges build up inside the cloud.

Answer these questions about the diagram above.

1. As a thunderstorm starts to form, updrafts carry __________________ upward in a cloud.

2. Heavy rain falls because __________________ and __________________ electric charges build up inside a cloud.

3. Describe how the electric charges are situated in a thunderstorm cloud.

____

____

4. What happens when the air moving upward rubs against the air moving downward?

____

____

Name ______________________ Date ____________

Interpret Illustrations

Lesson 8

How Do Hurricanes Form?

This is a cutaway diagram. Such diagrams are used to show both the inside and the outside of something. Here, arrows show you the paths of different types of winds and air masses in a hurricane. Other arrows show how the hurricane moves. Study these movements and read the labels.

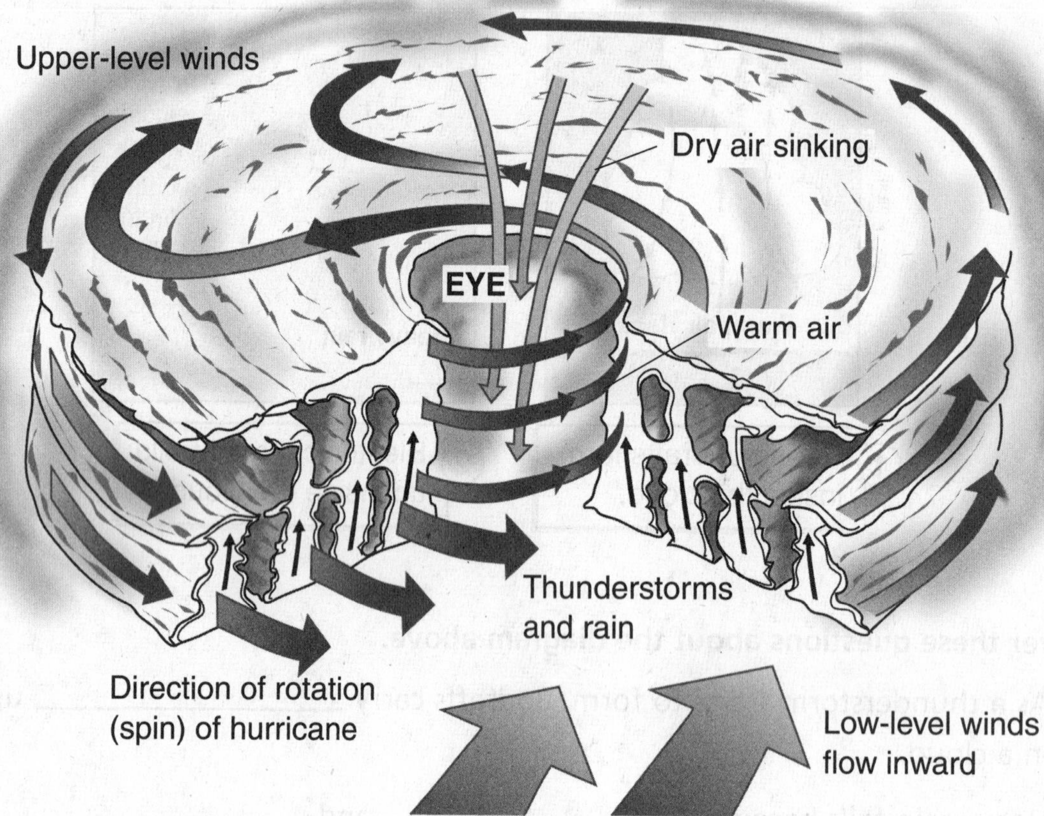

Answer these questions about the diagram above.

1. What forms at the center of the hurricane?

____

2. At the eye of the hurricane, ________________ is sinking.

3. A hurricane spins in a(n) ________________ direction.

4. Why do you think people are sometimes fooled by the eye of a hurricane?

____

____

____

Use with textbook page D78

Name ______________________ Date __________

Severe Storms

Vocabulary

thunderstorm
air pressure
cold front
tornado
storm surge
hurricane

Fill in the blanks.

1. A violent whirling wind that moves across the ground in a narrow path is called a(n) ________________.
2. Low air pressure in a storm can cause a(n) ________________, a bulge of water along the ocean shore.
3. A storm that forms over tropical oceans, near the equator is a(n) ________________.
4. The center of a hurricane is characterized by very low ________________.
5. The most common kind of severe storm is a ________________.
6. Thunderstorms occur most frequently when the weather is hot and a(n) ________________ approaches.

Answer each question.

7. How do strong updrafts contribute to the severity of a thunderstorm?

____

____

8. What is the difference between a hurricane and a tornado?

____

____

____

____

9. How can you protect yourself in a thunderstorm?

____

____

____

Use with textbook pages D74–D81

Name_______________________________ Date_____________

Cloze Test
Lesson 8

Severe Storms

Vocabulary

noise	rise	cold	rains
lightning	thunderstorms	updrafts	down

Fill in the blanks.

_______________ form in cumulonimbus clouds called thunderheads. Huge electric sparks called _______________ heat the air. This produces the _______________ named thunder. Thunderstorms usually feature strong winds and heavy _______________. Thunderstorms begin when strong _______________, or upward rushes of heated air, form inside clouds. Then the rising air rubs against the air going _______________, building up static electricity. When the weather is hot and humid with a(n) _______________ front approaching, thunderstorms are probable. These conditions cause the warm air to _______________ rapidly.

Name ______________________ Date ____________

Lesson Outline
Lesson 9

Climate

Fill in the blanks.  Reading Skill: **Sequence of Events** - questions 9, 12

What Is Climate?

1. The average weather pattern of a region over time is its ____________.
2. Two ways to describe a region's climate are:

   a. ______________________ and

   b. ______________________.

What Affects Climate?

3. Climate depends on factors that affect ____________ and ____________ over a period of time.
4. A measure of a place's distance north or south of the ____________ is the degrees of ____________.
5. The ______________________ at different latitudes causes differences in temperature.
6. The three climate zones that depend on latitude are

   a. ____________, near the equator;

   b. ____________, in the middle latitudes;

   c. ____________, at high latitudes near the North and South Poles.
7. Land areas heat up and cool off ____________ than water.

8. In winter, air temperatures over land are ______________ than temperatures over water.
9. Winds set up ______________ in the ocean, which move warm or cool air across the ocean surface.
10. Mountain ranges can block both cold air and ______________.

What Causes Climate Change?

11. Changes in the amount of energy the Sun sends out may be related to ______________.
12. Ocean currents help move Earth's heat, carrying heat from the ______________ to the ______________.
13. Dust and gas from volcanoes can block ______________ and cause cooling.

How Can Climate Affect You?

14. The main health problem in hot climates is ______________.

Climate

This diagram illustrates how different conditions of temperature and precipitation combine to produce many different types of climate. To understand this diagram, first read the labels in large type, which show the main factors in climate. Notice that the diagram is somewhat like a graph, showing how factors interact. Also study the drawings of different types of plants that grow in different climates.

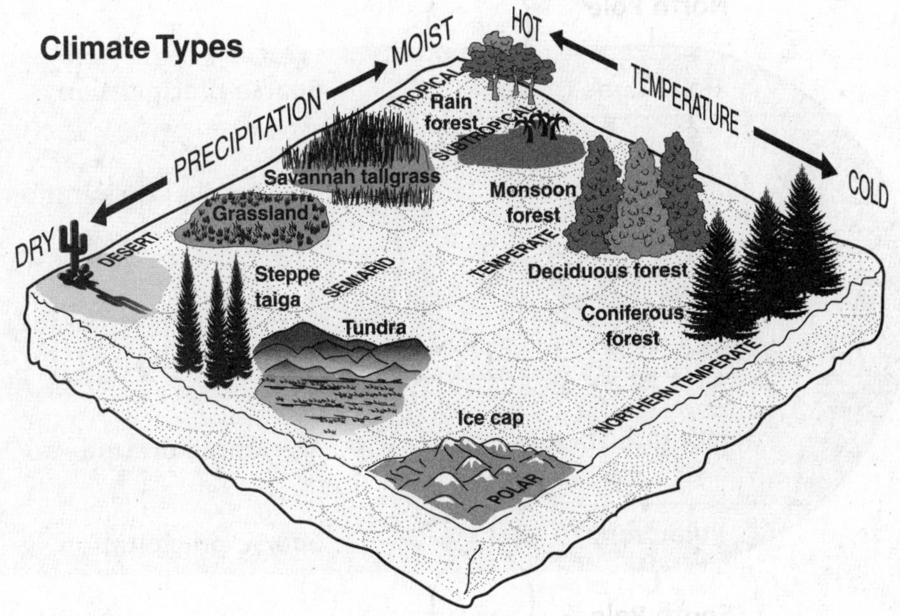

Answer these questions about the diagram above.

1. The lefthand side of the drawing shows the range of ______________ from dry to moist.
2. The righthand side of the drawing shows the range of ______________ from hot to cold.
3. At the corner with extreme heat and moisture, the climate is ______________.
4. Other hot climate types are ______________ and ______________.
5. What kind of forest is found in a climate that is both cold and moist? ______________.

Name______________________________ Date______________

Interpret Illustrations

Lesson 9

What Affects Climate?

Climate is affected by latitude, which is a measure of how far north or south of the equator a region is, and how much energy it gets from the Sun. The climate of a region is also affected by nearby bodies of water, and wind and ocean currents. All these forces help to distribute or concentrate the Sun's energy and determine how hot or cold a region may be.

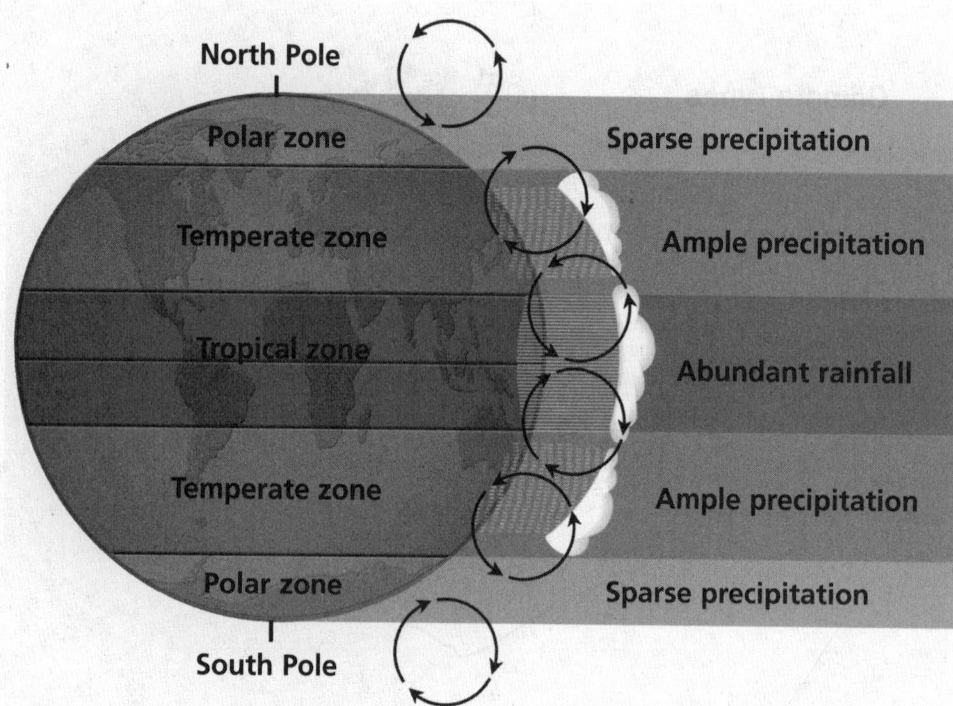

Answer these questions about the diagram above.

1. Would you expect to find a warmer climate in Earth's Tropical zone or its Temperature zone?

____

2. Would you expect there to be more rainfall in Earth's Polar zone or in its Temperate zone?

____

3. At the North Pole, do surface winds move north or south?

____

Name ______________________ Date ____________

Lesson Vocabulary

Lesson 9

Climate

Vocabulary

temperate
climate
latitude
temperature
equator

Fill in the blanks.

1. The average weather pattern of a region is its ______________.
2. The measure of how far north or south of the equator something is located is called its ______________.
3. It seems that during a sunspot maximum, the average ______________ of Earth increases slightly.
4. The eastern half of the United States is located in the ______________ climate zone.
5. Ocean currents carry heat away from the ______________ and the surrounding area.

Answer each question.

6. What information can the plants living in a region tell you about the region's climate? Why?

____

____

____

7. Why does an area's closeness to or distance from the ocean help describe that area's climate?

____

8. How does the atmosphere prevent us from getting too hot during the day?

____

____

Name ______________________ Date ____________

Cloze Test
Lesson 9

Climate

Vocabulary

climate	rain shadow	leeward	rainfall
global	sea level	windward	

Fill in the blanks.

Altitude is the measure of how high above ______________________ a place is. The higher a place is above sea level, the cooler its ______________________ is. Mountains affect ______________________ and climate. Air can be forced up a mountain by ______________________ wind patterns. As the air moves up the mountain, there is precipitation on the ______________________ side. Dry air moves down the ______________________ side of the mountain. A(n) ______________________ is the area where the dry air moves down the side of the mountain.

Name ______________________ Date ____________

Chapter Vocabulary

Chapter 11

Weather Patterns and Climate

Circle the letter of the best answer.

1. Which of these is NOT a factor considered in describing the climate of a region?
   a. distance from the ocean
   b. the number of mountains nearby
   c. the population of the region
   d. the amount of precipitation

2. A tornado
   a. has very high air pressure at its center.
   b. has winds less than 75 miles per hour.
   c. always travels in straight lines.
   d. has a very strong updraft at its center.

3. Doppler radar is used to spot
   a. airplanes landing and taking off.
   b. sunspots.
   c. the spinning movements of clouds.
   d. coastlines and ship traffic.

4. A storm surge is
   a. a period of heavier rain.
   b. a large swell in the ocean by the shore.
   c. a sudden increase in electrical power to your home.
   d. a tornado's speed.

5. When there are many dark spots on the Sun's surface it is a(n)
   a. sunset maximum.
   b. solar maximum.
   c. sunrise maximum.
   d. sunspot maximum.

6. A hurricane
   a. is formed over warm tropical waters.
   b. cannot be predicted.
   c. is a storm that follows a narrow path.
   d. only happens at the shoreline.

7. During a thunderstorm, you should
   a. stay outside until the storm passes.
   b. go outside and stand under the tallest tree.
   c. go to the nearest body of water.
   d. stay away from electrical wires and appliances.

8. A cold front occurs when cold air moves
   a. over cold air.
   b. under cold air.
   c. under warm air.
   d. over warm air.

9. Wearing loose clothing on hot days will
   a. make you sweat more.
   b. allow air to cool you.
   c. make you feel hotter.
   d. increase your body temperature.

10. A large area of the atmosphere where the air has similar properties is called a(n)
   a. water mass.
   b. land mass.
   c. air mass.
   d. solid mass.

11. Warm air moves in over a cold air mass in a(n)
   a. cold front.
   b. mild front.
   c. frigid front.
   d. warm front.

12. The boundary between two air masses with different temperatures is a(n)
   a. isobar.
   b. front.
   c. millibar.
   d. occluded front.

13. Thunderstorms form in
   a. cumulonimbus clouds.
   b. cirronimbulus clouds.
   c. cumulus clouds.
   d. cirrus clouds.

14. Which kind of air mass has cold, dry air?
   a. Maritime Polar
   b. Continental Polar
   c. Maritime Tropical
   d. Continental Tropical

Name That Word

Read each description, then write the word or words being described. Use the Word Box to check your spelling.

Word Box				
front	storm surge	cirrus cloud	thunderstorm	air pressure
condensation	evaporation	stratus cloud	humidity	Coriolis effect

1. These words identify a cloud that forms in a blanketlike layer.

   ____

2. These words identify a feathery high-altitude cloud made of ice crystals.

   ____

3. This word identifies the changing of gas into a liquid.

   ____

4. This word identifies the changing of liquid into a gas.

   ____

5. This word identifies the amount of water vapor in the air.

   ____

6. These words identify an object's curving movement caused by Earth's rotation.

   ____

7. These words identify the force put on an area by the weight of the air above it.

   ____

8. This word identifies a boundary between air masses with different temperatures.

   ____

9. This word identifies the most common severe storm, formed in cumulonimbus clouds. ______________________________

10. These words identify a great rise of the sea along a shore caused by low-pressure clouds. ______________________________

Name ______________________ Date ____________

Unit Vocabulary

Unit D

Crossword

Read each clue. Write the answer.

Across

1. Change from a gas to a liquid
2. Weather map line connecting places with equal pressure
3. Pellets of ice
5. Layers of gas surrounding Earth
7. Opposite of cool
12. Boundary between air masses of different temperatures
13. Measures air pressure
14. Change from a liquid to a gas

Down

1. A region's average weather pattern
4. Large area in the atmosphere where the air has similar properties
6. Amount of water vapor in the air
8. Funnel-shaped wind storm
9. Moving air
10. What the lower atmosphere is like at a given time
11. A cloud at ground level

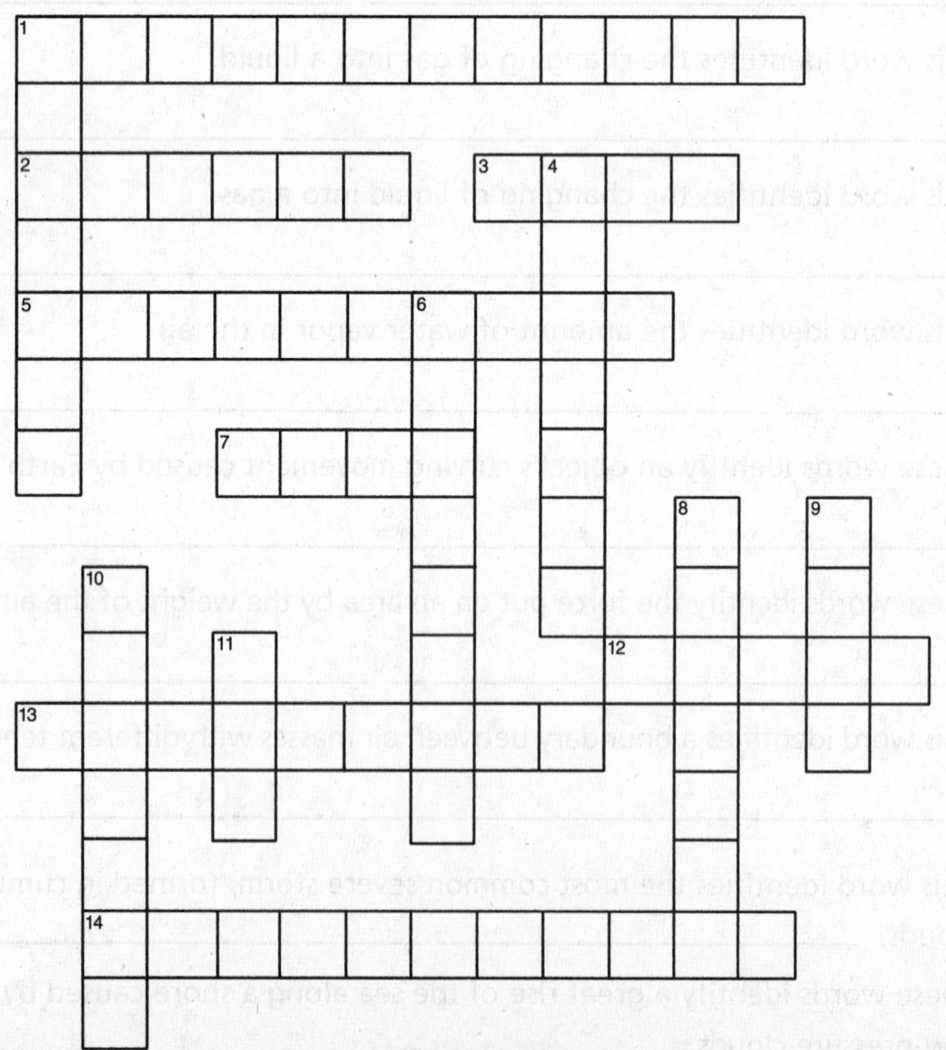

Crack-a-Code

Code Key

A B C D E F G H I J K L M N O P Q R S T U V W X Y Z

Use the Code Key to help you decode each word. Then draw a line to its meaning.

1. ______________ a. the amount of water vapor in the air

2. ______________ b. average weather in a region

3. ______________ c. device to measure wind speed

4. ______________ d. cumulonimbus cloud in which a thunderstorm forms

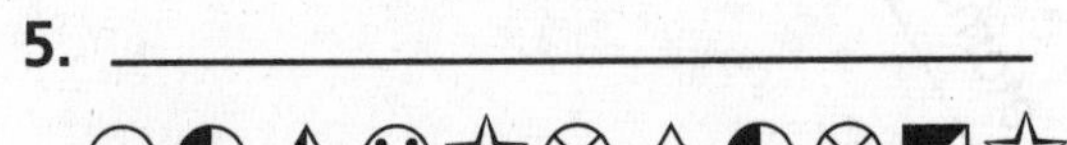

5. ______________ e. weather map line connecting places with equal air pressure

6. ______________ f. what the lower atmosphere is like at a given place and time

7. ______________ g. pellets of ice

Name ______________________ Date __________

Chapter Graphic Organizer

Chapter 12

Properties and Structure of Matter

You can use a spider map to organize the main ideas and supporting details of a topic such as properties of matter. Look at the example shown below. The central oval contains the topic being studied. Each branch extending from the oval represents a main idea of the topic. Each branch extending from one of the main ideas represents a supporting detail.

Example

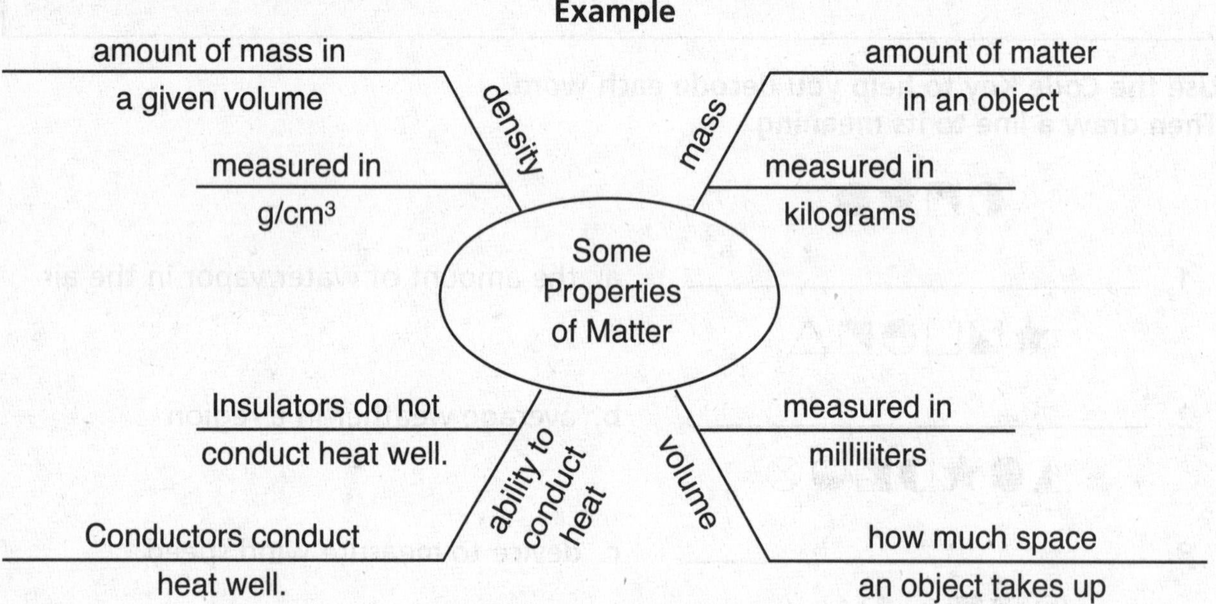

Complete the spider diagram below for the topic What Matter Is Made Of. The main ideas have already been filled in for you.

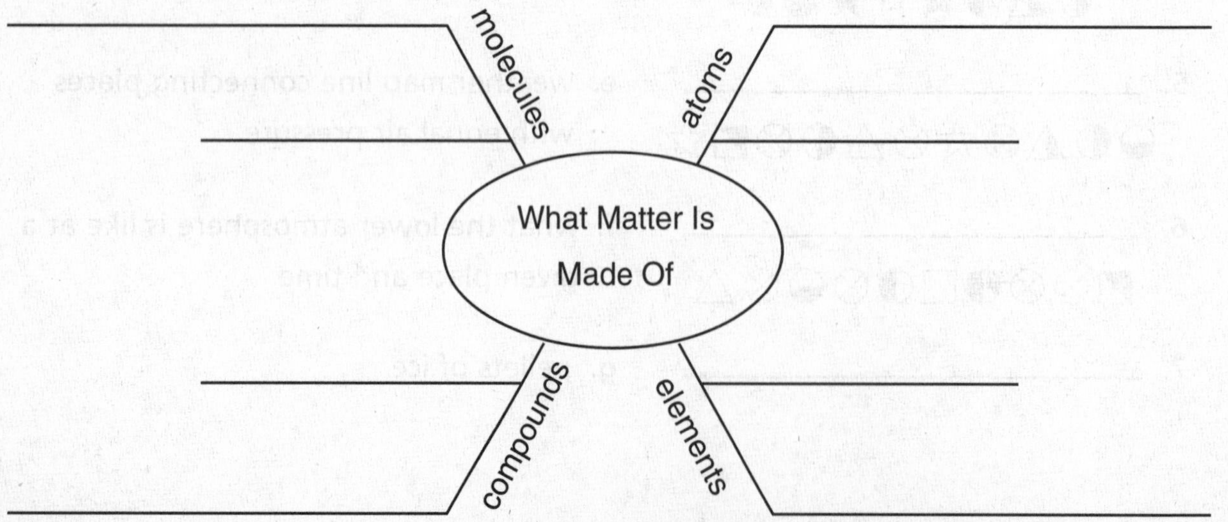

Main Idea and Supporting Details

The main idea tells what a story's about. Other sentences contain details that support, or tell more about, the main idea.

Read the following paragraphs, based on information from this chapter of your textbook. **Circle the main idea and underline the supporting details.**

1. What makes things sink or float? It all depends on density, or how much matter an object has for its size. The object's mass divided by its volume gives us its density. If the object's density is less than water's density, the object floats in water. If the object's density is greater than water's density, the object sinks.
2. Atoms contain three kinds of particles—protons, neutrons, and electrons. The protons and neutrons are in the nucleus, the atom's very dense center. The electrons are outside the nucleus. Electrons have about 2,000 times less mass than protons or neutrons. Protons carry one positive electric charge, electrons carry one negative electric charge, and neutrons have no charge at all!

Now reread the Science Magazine article "Animals: Icy Survival." **Then write the main idea of the article, and list at least four supporting details.**

Main Idea: ______________________________

____

Supporting Details:

1. ______________________________

2. ______________________________

3. ______________________________

4. ______________________________

Name ______________________ Date ____________

Chapter Reading Skill

Chapter 12

The Very Idea!

Read the following paragraphs, based on information in this chapter of your textbook. One paragraph is based on the Science Magazine article "Animals: Icy Survival." **After reading the paragraphs, write the main idea of each and list some of the supporting details.**

1. Dewdrops are not raindrops. Where do the dewdrops on a flower come from? They are water from the air around the flower! In the air the water was a gas, or vapor. When the surface of the flower became cold enough, the water vapor near the flower condensed. It became liquid water. The condensed water formed dewdrops on the petals of the flower.

Main Idea:

____

____

Supporting Details:

____

____

____

____

____

2. Why does ice float? Doesn't water become denser when it turns to ice? To answer that, you have to look at what happens to water molecules when they get cold enough to freeze. As water freezes, its molecules are kept farther apart than when water is liquid. Ice is only nine-tenths as dense as liquid water, so when water freezes, it expands. That's why ice floats!

Main Idea:

____

____

Supporting Details:

____

____

____

____

____

Use with textbook pages E2–E47

Name ______________________ Date ____________

Lesson Outline

Lesson 1

Physical Properties

Fill in the blanks.  Reading Skill: **Main Idea and Supporting Details** - questions 13, 18

What Is Matter?

1. A measure of the amount of matter in an object is ______________.
2. A(n) ______________ measures mass.
3. Mass is often measured in ______________.
4. A measure of how much space a sample of matter takes up is ______________.
5. Anything that has mass and takes up space is ______________.
6. The ______________ of an object is a measure of the force of gravity between Earth and the object.
7. An object's ______________ is a measure of the amount of matter in the object compared to known masses.
8. Scientists use a quantity called the ______________ to measure force.
9. An object's weight depends on its ______________ in the universe.

What Is Density?

10. The ______________ of an object tells us how massive something is for its size.
11. As long as conditions such as ______________ do not change, the density of a substance does not change.
12. Each ______________ has its own density.

How Dense Are Solids, Liquids, and Gases?

13. The property called ______________ is a measure of how tightly packed matter is.
14. As more matter gets packed into the same amount of space, the material's density ______________.

Name ______________________ Date ____________

How Does Density Make Things Sink or Float?

15. Air is much less dense than water, so a beachball ____________.

16. An object's ability to float is called its ____________.

17. Objects have enough buoyancy to float when they are less ____________ than the liquid in which they are placed.

What Are Conductors and Insulators?

18. When materials ____________ energy well, they allow energy to flow through them easily.

19. When materials ____________ against the passage of energy, they do not readily permit energy to flow.

20. Metals like copper are good ____________ of electricity.

What Is Magnetism?

21. A(n) ____________ has north and south poles.

22. Opposite poles ____________ one another.

How Do We Use Properties of Matter?

23. Engineers and scientists use ____________ of matter when they design and build things.

24. Materials with very low density and relatively great strength, called ____________, are very good insulators against heat.

Use with textbook pages E4–E17

Name ______________________ Date ____________

Interpret Illustrations
Lesson 1

What Is Matter?

These two drawings compare mass and weight on Earth and on the Moon. Think about what is happening to the astronaut each time. Note what is being measured, how it is measured, and what units are used.

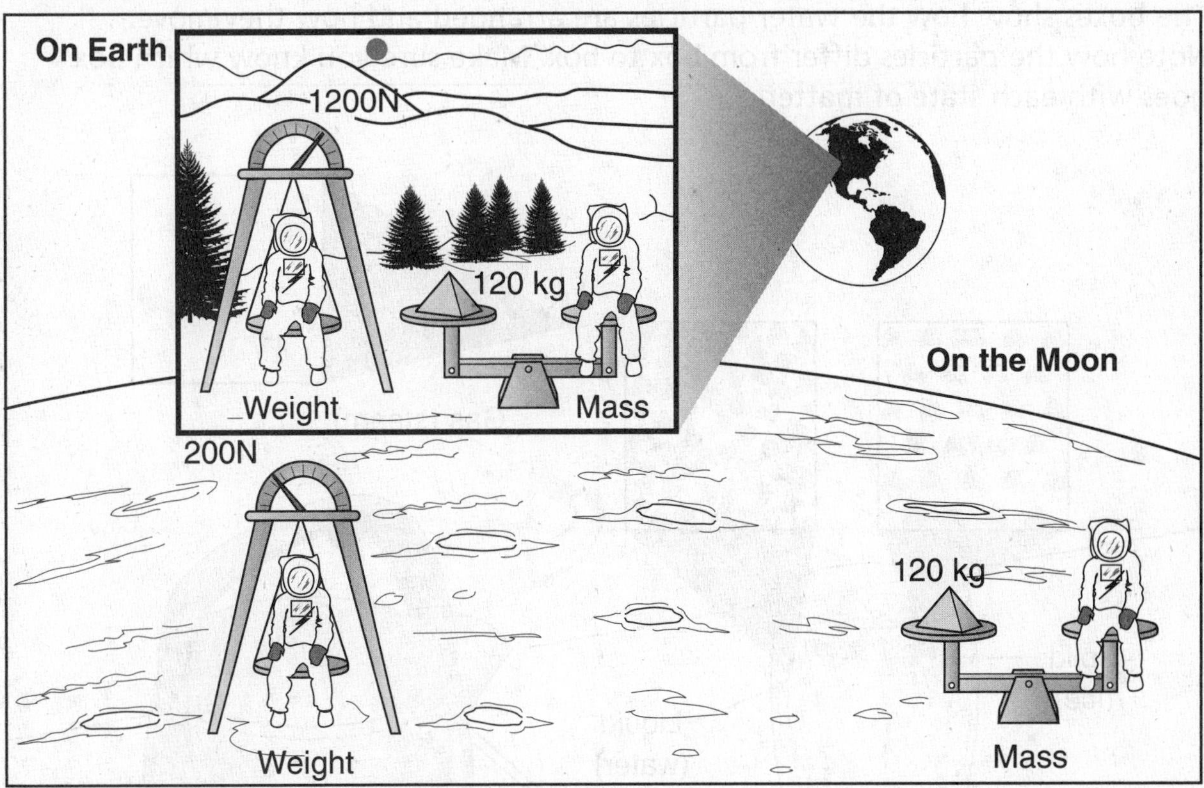

Answer these questions about the diagram above.

1. What is being measured by the pan scale in each drawing? What unit is used?

____

2. What is being measured on the double-pan balance? What unit is used?

____

3. What happens to the weight of the astronaut on the Moon?

____

4. What happens to the mass of the astronaut on the Moon?

____

Name ______________________ Date ______________

Interpret Illustrations

Lesson 1

How Dense Are Solids, Liquids, and Gases?

These drawings show three different states of matter, using water as an example. The boxes show how the water particles are arranged and how they move. Note how the particles differ from box to box. Make sure you know which box goes with each state of matter.

Answer these questions about the diagram above.

1. Which particles are the most orderly? ______________
2. Which particles are moving the most? ______________
3. Which particles are close together but can slide around each other?

____

Name ______________________ Date __________

Lesson Vocabulary

Lesson 1

Physical Properties

Match the correct letter with the description.

_____ 1. to allow energy or electricity to flow through readily

_____ 2. a measure of the force of gravity between Earth and an object

_____ 3. matter in which particles line up pole to pole

_____ 4. the amount of space an object takes up

_____ 5. all of the gases, liquids, and solids in the world

_____ 6. the amount of matter in an object

_____ 7. not to allow energy to flow through readily

_____ 8. tells us how massive something is for its size

Vocabulary

a. mass
b. volume
c. weight
d. density
e. magnetic
f. conduct
g. insulate
h. matter

Use the terms above to identify the property used.

9. use a flashlight ______________

10. use a compass ______________

11. use a paperweight ______________

12. sink marbles in the bottom of a fish bowl ______________

13. drink 8 ounces of water ______________

14. wear layers during the winter ______________

15. step on a scale ______________

Name ____________________________ Date ______________

Cloze Test

Lesson 1

Physical Properties

Vocabulary

conduct	measure	insulate
dense	float	metal
handles	mass	
volume	properties	

Fill in the blanks.

To tell one type of matter in objects from another you can ____________________ their ____________________. The amount of matter in an object is its ____________________. The amount of space it takes up is its ____________________. If one liquid is less ____________________ than another it will ____________________ on top of it. Materials that ____________________ energy allow it to flow through readily. Cooking pans are made of ____________________ since it lets heat pass through. However, the ____________________ are made of wood or ceramic so your hands don't get burned. These materials ____________________ against the passage of energy.

Use with textbook pages E4–E17

Name ____________________ Date __________

Lesson Outline
Lesson 2

Elements and Compounds

Fill in the blanks.  Reading Skill: **Main Idea and Supporting Details** - questions 1, 14

What Is Matter Made Of?

1. The basic building blocks of all matter are ____________.
2. Elements are pure substances that ____________ be broken down into simpler substances.

What Are Compounds?

3. Any substance that is formed by the ____________ of two or more elements is called a compound.
4. Compounds can only be broken apart by ____________.
5. Compounds have ____________ properties than the elements that make them up.

How Do You Write a Compound's Name?

6. Scientists write symbols for compounds called ____________.
7. A formula contains numbers below the elements called ____________.
8. The subscripts in chemical formulas tell us the ____________ of particles that combine in a compound.

What Are Elements Made Of?

9. A(n) ____________ is the smallest unit of an element that retains the properties of the element.

What Is Inside Atoms?

10. The three kinds of particles that atoms contain are ____________, ____________ and ____________.
11. The tiny, very dense body in the atom's center is called the atomic ____________.
12. The number of ____________ in an atom determines what element it is.

Name ______________________ Date ____________

Lesson Outline
Lesson 2

What Properties Do Elements Have?

13. We now know of ____________ elements.
14. Some elements take part in chemical ____________ much more easily than others.
15. About three-fourths of the elements are ____________.
16. Properties of metals include:
    a. ______________________,
    b. ______________________, and
    c. ______________________.

How Can the Elements Be Grouped?

17. In 1869, Mendeleyev found that the properties of the elements went through ____________.
18. The groups in Mendeleyev's table contained elements with similar ____________ properties.
19. We call Mendeleyev's table the periodic table after the periodic changes found in the elements' ____________.

What Are Molecules?

20. Particles that contain more than one atom joined together are called ____________.
21. Molecules of ____________ always contain only one kind of atom.
22. Molecules that have different kinds of atoms joined together are called ____________.
23. When a compound forms from elements, changes occur in the way that the atoms are ____________.

How Do We Use Compounds?

24. Petroleum is a complex mixture of ____________.
25. A fuel we use in cars is called ____________.

Name ______________________ Date __________

Interpret Illustrations

Lesson 2

What Is Inside Atoms?

These three pictures show three very common atoms: hydrogen, helium, and carbon. Note how each proton, neutron, and electron is represented. Then count the number of each kind of these particles in each atom.

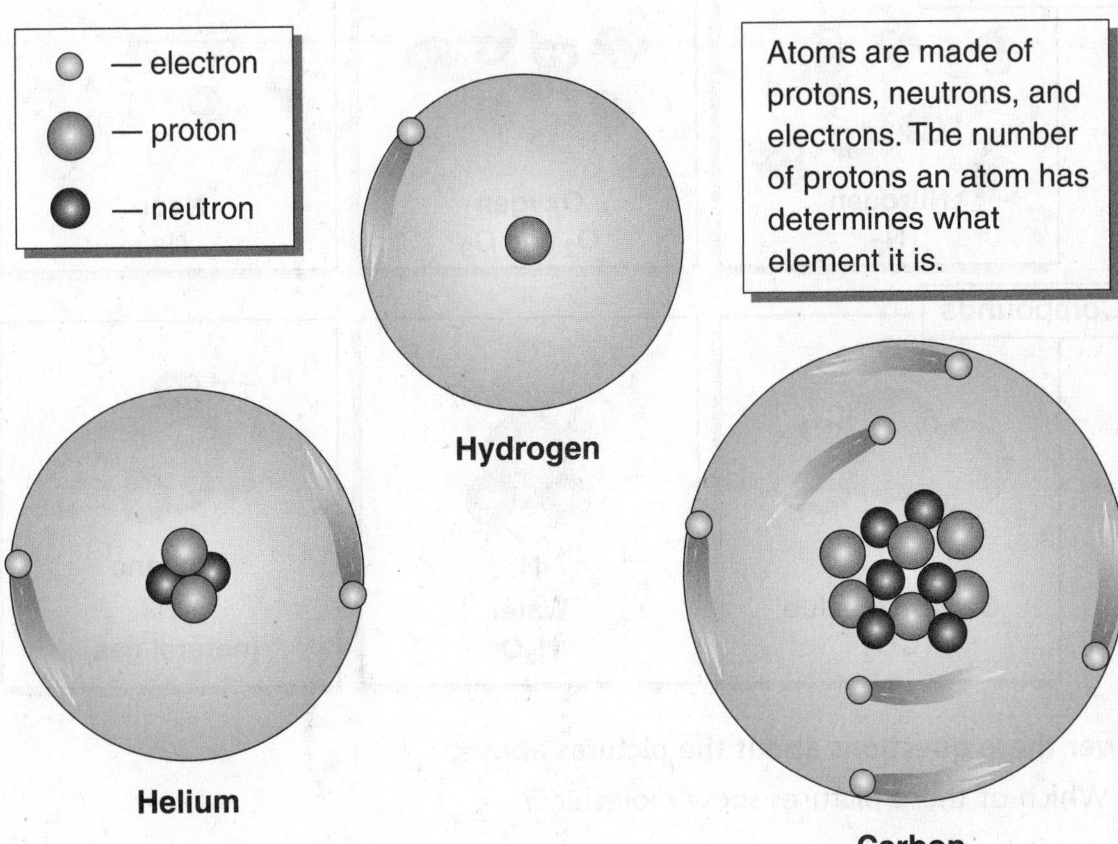

— electron
— proton
— neutron
Atoms are made of protons, neutrons, and electrons. The number of protons an atom has determines what element it is.
Hydrogen
Helium
Carbon

Answer these questions about the diagram above.

1. Which two particles are found in each of the atoms?

   ____

2. Which particle is NOT found in Hydrogen? __________

3. Which particle is NOT found in the nucleus? __________

4. Which atom weighs the most? Why?

   ____

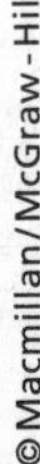

Name_______________________ Date___________

Interpret Illustrations

Lesson 2

What Are Molecules?

These pictures show different elements and compounds. Compare the picture of each with the formula below it.

Elements

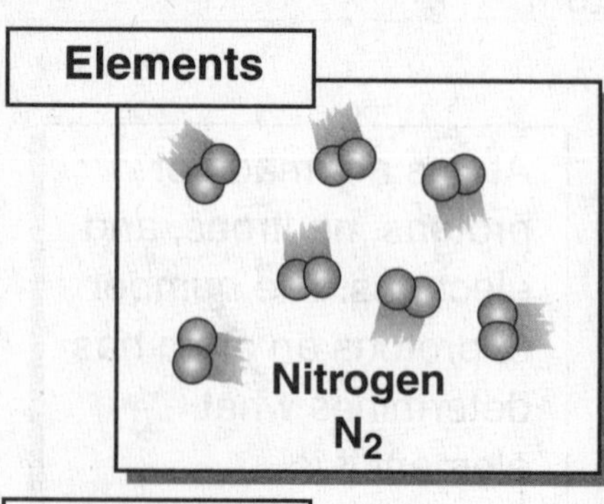
Nitrogen
N_2

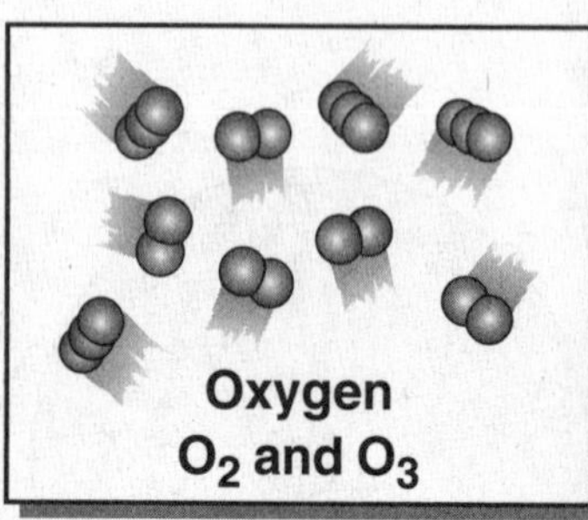
Oxygen
O_2 and O_3

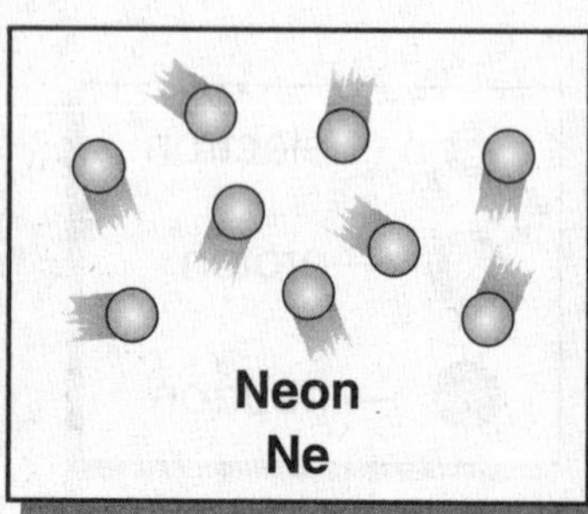
Neon
Ne

Compounds

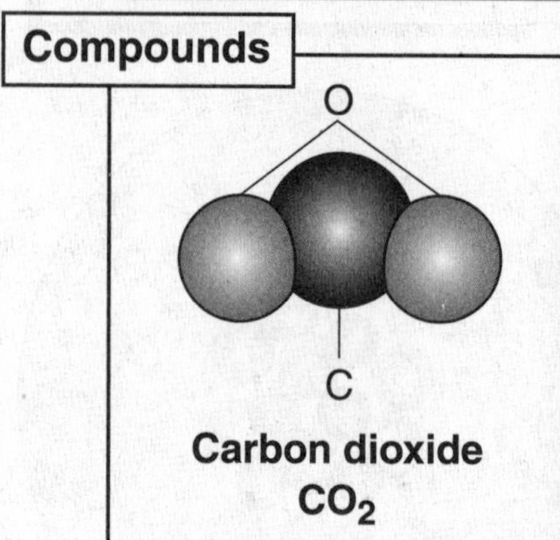

Carbon dioxide
CO_2

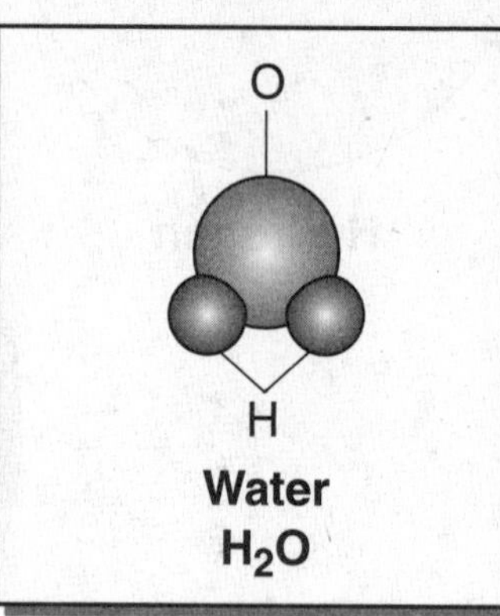

Water
H_2O

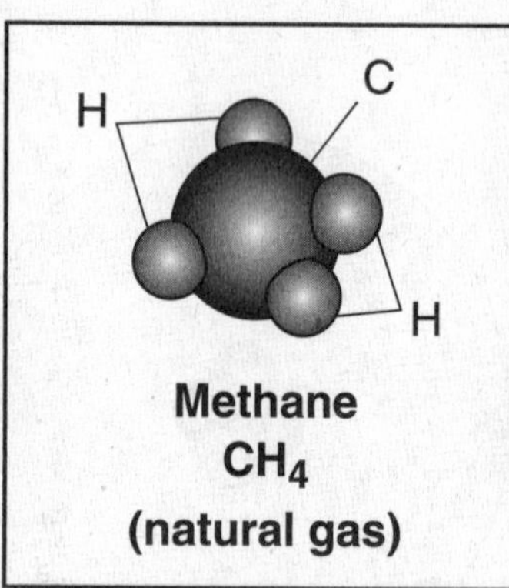

Methane
CH_4
(natural gas)

Answer these questions about the pictures above.

1. Which of these pictures show molecules?

____

2. What kinds of atoms make up these molecules?

____

3. What is the ratio of the number of oxygen atoms to the number of carbon atoms in the carbon dioxide molecule? ____________

4. How are the elements nitrogen and oxygen different from neon?

____

____

5. How do the pictures compare to the chemical formulas?

____

Name____________________ Date__________

Lesson Vocabulary
Lesson 2

Elements and Compounds

Match the correct letter with the description.

_____ **1.** the dense center part of an atom

_____ **2.** a chemical combination of two or more elements into a single substance

_____ **3.** the smallest unit of an element that still has the properties of the element

_____ **4.** contains the symbols for the elements that make it up

_____ **5.** an uncharged particle in the nucleus of an atom

_____ **6.** a particle that contains more than one atom joined together

_____ **7.** a basic building block of matter

_____ **8.** a particle with a negative charge moving around the nucleus of an atom

_____ **9.** one or two letters that identify an element

_____ **10.** a particle with a positive charge in the nucleus of an atom

Vocabulary

- **a. element**
- **b. compound**
- **c. atom**
- **d. proton**
- **e. neutron**
- **f. electron**
- **g. nucleus**
- **h. molecule**
- **i. chemical formula**
- **j. symbol**

Answer each question.

11. Which elements are poor conductors of electricity? ____________________

____

12. On the periodic table, where can you find the metalloids?

____

13. What is the difference between an atom and a molecule?

____

____

Name ______________________ Date __________

Cloze Test

Lesson 2

Elements and Compounds

Vocabulary

electron	compound	elements
symbol	properties	atom
table	neutron	proton

Fill in the blanks.

Scientists have identified 112 ________________. The smallest unit of each of these pure substances that has its properties is a(n) ________________. These units are made of smaller particles. The ________________ has a negative charge, the ________________ has a positive charge, and the ________________ has no charge. Each substance is given a special ________________ of one or two letters. A scientist named Dmitri Mendeleyev created a periodic ________________. He grouped the substances according to chemical ________________. Two or more substances can combine chemically to form a new ________________.

Name ______________________ Date __________

Lesson Outline
Lesson 3

Solids, Liquids, and Gases

Fill in the blanks.  Reading Skill: **Main Idea and Supporting Details** - questions 2, 4, 6, 14, 15, 16

What Are the States of Matter?

1. When water absorbs enough heat, it turns into a(n) __________ called steam.
2. The three states of matter are:
   a. __________,
   b. __________, and
   c. __________.
3. When a change of __________ occurs, the identity of the substance stays the same.
4. Substances change state because their __________ are arranged in a different way.
5. The molecules of any substance are __________ to each other.
6. Adding or removing __________ makes substances change from one state to another.
7. When molecules are linked in organized positions, a(n) __________ results.
8. When heat is absorbed by a solid, the molecules vibrate __________.
9. Heat causes the molecules of a solid to __________ from each other, making the solid become a liquid.
10. Every substance has its own particular melting point and __________.
11. When a substance is melting or boiling, its __________ stays the same.
12. When heat is removed from a boiling substance, it __________.
13. The melting point is also called the __________ point.

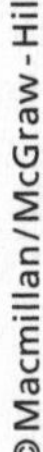

Name_______________________________ Date_____________

Lesson Outline
Lesson 3

How Can Matter Change to a Gas?

14. When a liquid ________________, it gradually changes to a gas.
15. When a liquid ________________, it changes to a gas rapidly.
16. Once all the liquid becomes steam, the ________________ goes up.

What Are the Properties of Solids, Liquids, and Gases?

17. Solids keep their ________________.
18. Matter that takes the shape of their containers are ________________.
19. Gases fill the ________________ of their containers.
20. When the temperature of a material increases, its particles move ________________.
21. Materials ________________ as they get hotter.
22. Materials ________________ as they get cooler.
23. Substances that are ________________ expand or contract the most with changing temperature.

How Can Expansion and Contraction Be Used?

24. The liquid in a ________________ expands or contracts with changes in temperature.

Name_______________________________ Date_______________

Interpret Illustrations
Lesson 3

What Are the States of Matter?

This diagram shows what happens to particles during changes of state. This is not what you would actually see, but the diagram helps you understand what is actually happening. Focus on how much the particles move.

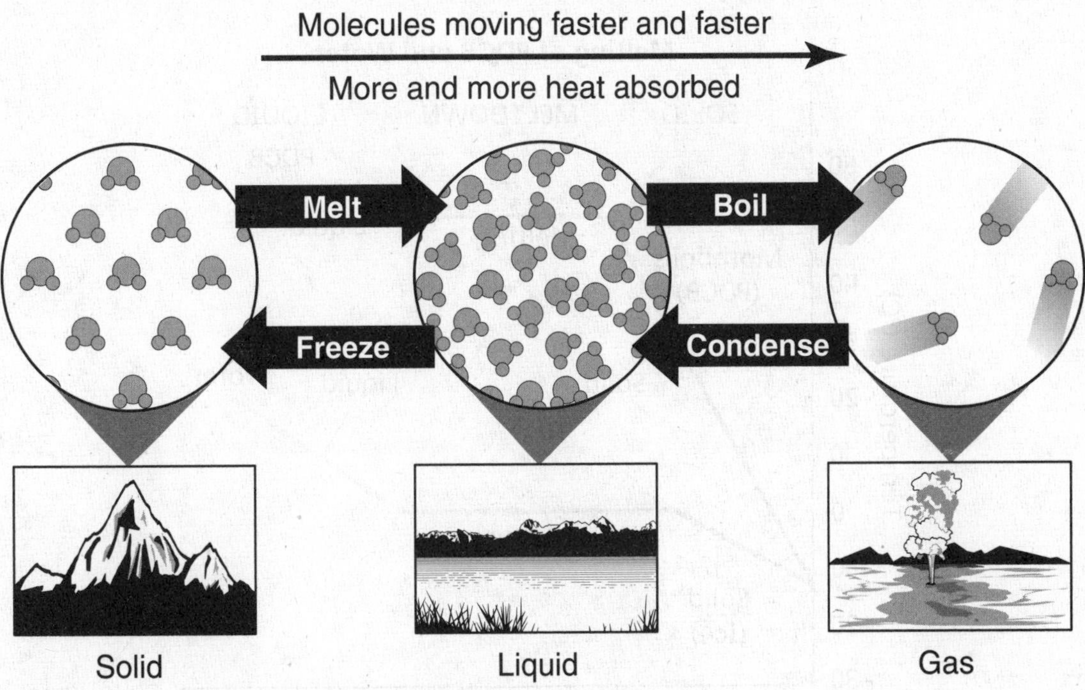

Answer these questions about the diagram above.

1. What happens to the speed of molecules as they absorb heat?

____

2. In which state have the particles absorbed the most heat? _______________

3. Which particles show the least amount of motion? _______________

4. In which state do the particles occupy the greatest amount of space?

____

Name ______________________ Date ____________

Interpret Illustrations

Lesson 3

What Are the States of Matter?

This line graph shows the change in temperature of water and PDCB over time. To get the data, a sample of each is heated and its temperature at each time interval is recorded. As you study a graph, note the title, the two axes, and the general shape of the lines.

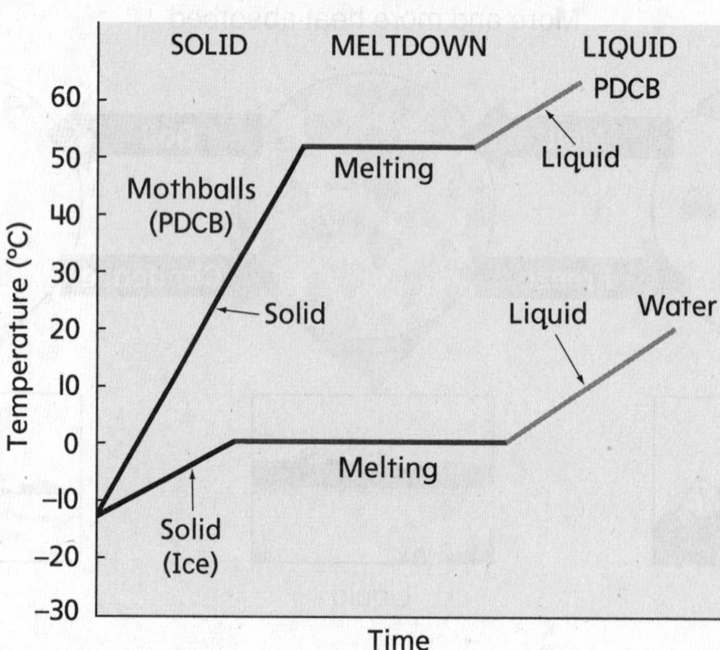

Answer these questions about the diagram above.

1. If water has a melting point of 0°C, which line shows the data for water?

____

2. What is the melting point of PDCB? ______________

3. Why does the water line flatten out after 0°C?

____

4. What do you think is happening during the time the temperature is not increasing? ____

5. What state of matter is the PDCB in until 52°C? ______________

Name ____________________ Date __________

Lesson Vocabulary
Lesson 3

Solids, Liquids, and Gases

Vocabulary

- a. state of matter
- b. melting point
- c. boiling point
- d. freezing point
- e. evaporation
- f. condensation
- g. change of state
- h. expand
- i. contract

Match the correct letter with the description.

_____ 1. the temperature at which a liquid changes state into a solid

_____ 2. the temperature at which a solid changes state into a liquid

_____ 3. a liquid slowly changing into a gas

_____ 4. any of the forms matter can exist in

_____ 5. a gas turning to a liquid

_____ 6. the temperature at which a liquid changes state into a gas

_____ 7. the temperature stays constant during this even though heat is being added or removed

_____ 8. cooling causes particles to get closer together and materials do this

_____ 9. materials spread out as they get hotter

10. Explain why the temperature of water stays the same for a while once it starts boiling.

____

____

11. Name one way you benefit from each of the following states of matter.

solid _______________________________

liquid _______________________________

gas _______________________________

12. How are boiling and evaporation the same? How are they different?

____

____

____

Name ______________________________ Date ____________

Cloze Test

Lesson 3

Solids, Liquids, and Gases

Vocabulary

solid	expand	evaporation
freezing	properties	liquid
gas	states	
molecules	contract	

Fill in the blanks.

Most substances exist in one of three __________________ of matter. Whenever a change occurs, the substance gains new __________________. This happens because the __________________ are arranged differently. Each pure substance has a melting point, or temperature at which a(n) __________________ changes into a(n) __________________. Each also has a boiling point at which the liquid changes into a(n) __________________. The process by which a liquid slowly turns into a gas is known as __________________. The temperature at which a liquid changes into a solid is known as the __________________point. Most materials __________________, or become smaller, as they get cooler. They __________________, or become larger, as they get hotter.

Name ______________________ Date __________

Chapter Vocabulary

Chapter 12

Properties and Structure of Matter

Circle the letter of the best answer.

1. A particle outside the nucleus of an atom is a(n)
   - a. proton.
   - b. electron.
   - c. insulate.
   - d. molecule.

2. The amount of space an object takes up is its
   - a. density.
   - b. mass.
   - c. volume.
   - d. weight.

3. A material that does not readily permit heat to flow through it is a material that
   - a. insulates.
   - b. conducts.
   - c. measures.
   - d. weighs.

4. The amount of mass contained in a given volume is its
   - a. buoyancy.
   - b. density.
   - c. insulation.
   - d. weight.

5. A measure of the force of gravity between Earth and an object is its
   - a. density.
   - b. mass.
   - c. volume.
   - d. weight.

6. The amount of matter in an object is called its
   - a. density.
   - b. mass.
   - c. volume.
   - d. weight.

7. A charged particle in the nucleus of the atom is a(n)
   - a. electron.
   - b. molecule.
   - c. proton.
   - d. neutron.

8. A chemical combination of two or more elements into a single substance is a(n)
   - **a.** compound.
   - **b.** metal.
   - **c.** molecule.
   - **d.** nucleus.

9. A group of more than one atom joined together that acts like a single particle is a(n)
   - **a.** neutron.
   - **b.** metal.
   - **c.** molecule.
   - **d.** nucleus.

10. A pure substance that cannot be broken down into anything simpler is a(n)
   - **a.** compound.
   - **b.** element.
   - **c.** molecule.
   - **d.** nucleus.

11. The smallest unit of an element that still has the properties of the element is a(n)
   - **a.** atom.
   - **b.** compound.
   - **c.** molecule.
   - **d.** nucleus.

12. The dense center part of an atom is its
   - **a.** molecule.
   - **b.** neutron.
   - **c.** nucleus.
   - **d.** proton.

13. Any of the forms matter can exist in is what is called a(n)
   - **a.** change of matter.
   - **b.** molecule.
   - **c.** periodic.
   - **d.** state of matter.

14. Each different substance has its own boiling temperature called the
   - **a.** melting point.
   - **b.** freezing point.
   - **c.** liquefying point.
   - **d.** boiling point.

Chapter Summary

1. What are four vocabulary words you learned in the chapter? Write a definition for each.

2. Which diagram best describes a main idea in this chapter?

3. What are two main ideas that you learned in this chapter?

Name_________________________ Date_____________

Chapter Graphic Organizer
Chapter 13

Forms of Matter and Energy

You can use a diagram, like the one below about the movement of heat, to show how one or more events may cause another event to happen. Each arrow points from a cause to an effect. Notice that the effect in this diagram can be caused by three different events.

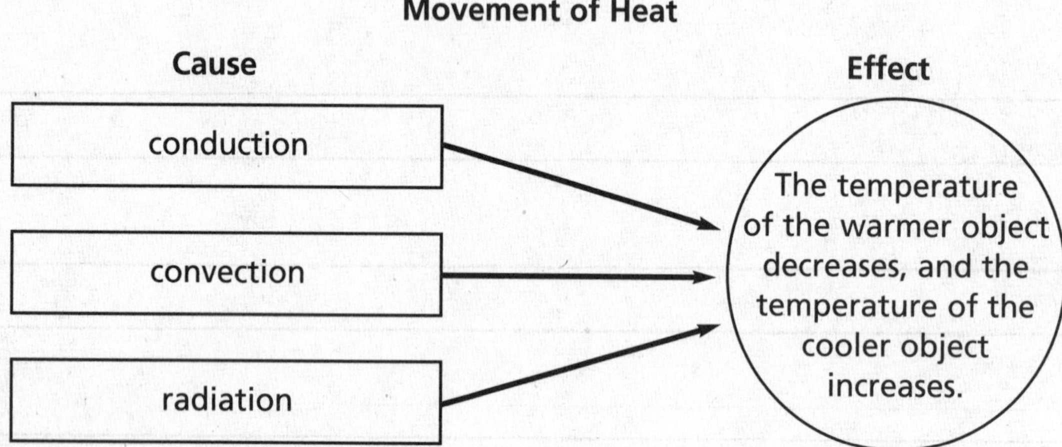

Complete the cause-and-effect diagram shown below. This diagram involves the separation of a mixture of sand, salt, and iron filings. Notice that each effect in this diagram has only one cause.

Separation of a Mixture

	Cause		Effect
1.	A magnet is held near the mixture.	→	
2.		→	
3.		→	
4.		→	The water boils away, leaving the salt behind.

Name ______________________ Date ____________

Cause and Effect

Often there's more than one cause for an effect. For example, you toss a toy bone for your dog to catch. The dog jumps, but misses the toy, and bumps into your mom's favorite potted plant. The plant falls over, and mud spills all over the rug! You, your dog, and the toy bone are all causes of the mess . . . but only you can clean it up!

Read each cause below. Write two possible effects.

1. Our school bus had a flat tire . . .

____

____

2. Jamie forgot to set his alarm . . .

____

____

3. All the fifth graders in our school took the test . . .

____

____

4. Carmen found a CD player that someone left in the park . . .

____

____

5. Brian forgot to lock the front door when he came home . . .

____

____

6. Alex and Andy saved their allowances all year . . .

____

____

Name ______________________ Date __________

Chapter Reading Skill
Chapter 13

If . . . Then Fun

Look at the drawing below. Then read each **If . . . then** statement and follow directions.

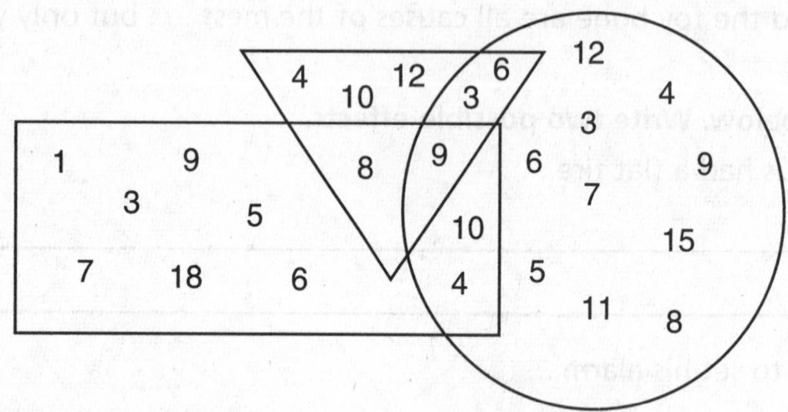

1. **If** the numbers are in the triangle **only, then** add them.
2. **If** the numbers are in the circle **only, then** add them.
3. **If** the numbers are in the rectangle **only, then** add them.
4. **If** the numbers are in the triangle **and** rectangle **only, then** add them.
5. **If** the numbers are in the triangle **and** circle **only, then** add them.
6. **If** the numbers are in the rectangle **and** circle **only, then** add them.
7. **If** the numbers are in the triangle, circle, **and** rectangle, **then** add them.
8. **If** there are numbers in boxes 1 through 7, **then** add them together.
9. **If** the answer in box 8 is 192 or more, **then** write a 5 in this box.

Name ______________________ Date ____________

Lesson Outline
Lesson 4

Mixtures and Solutions

Fill in the blanks.  Reading Skill: **Cause and Effect** - questions 5, 15, 17

What Is a Mixture?

1. Materials that are physically combined form ____________.
2. Materials that are chemically combined form ____________.
3. In mixtures, the parts keep their original ____________.
4. A compound has ____________ properties from the substances that combine to make it.

How Can Mixtures Be Classified?

5. If the substances in a mixture blend completely, we call this a(n) ____________.
6. Mixtures that have properties in between those of solutions and heterogeneous mixtures are called ____________.

What Kinds of Solutions Are There?

7. Solutions never settle into ____________.
8. The substance that makes up the smaller part of a solution is called the ____________.
9. The substance that makes up the larger amount of the solution is the ____________.

What Is Solubility?

10. Solubility explains the ability of a solute to ____________ in a solvent.
11. Temperature and ____________ affect solubility.
12. The concentration of a solution is a ____________ of the amount of solute that is dissolved in a solvent.

What Are Heterogeneous Mixtures Like?

13. Rocks are examples of ______________ mixtures.
14. Heterogeneous mixtures can settle into ______________.
15. When one substance is insoluble, or does not dissolve in a solvent, ______________ are formed.

What Types of Colloids Are There?

16. Milk is an example of a(n) ______________.
17. Colloids that have either liquid drops or solid particles spread through a gas are ______________.

How Can Mixtures Be Separated?

18. A(n) ______________ separation gets the parts of a mixture away from one another without changing their identities.
19. To separate sand and wood chips, first pour in ______________.

Which Resources Come from Mixtures?

20. Seawater is a mixture that can be used as a(n) ______________ for fresh water.
21. Crude oil, our major ______________ source, is a mixture, too.

Name ______________________ Date ____________

Interpret Illustrations

Lesson 4

How Can Mixtures Be Separated?

These drawings show someone separating a mixture of salt and sand. Note what is happening in each step and think about the processes that are going on.

1
To separate a mixture of sand and salt, pour in water and stir. The salt dissolves, but the sand doesn't.
2
Use a filter to separate the sand from the salt water.
3
Then let the water evaporate to get back the salt.

Answer these questions about the diagram above.

1. What will happen after water is added to the salt and sand mixture?

   ____

2. In the second picture, what does the filter paper keep out of the second beaker? ____________________

3. What will you find in the bottom of the second beaker in the second drawing?

   ____

4. In the third picture, what is leaving the beaker? ____________________

5. What is left behind in the beaker in the third picture? ____________________

Name ______________________ Date ____________

Interpret Illustrations

Lesson 4

Separating Alcohol and Water

This diagram shows a distillation apparatus that is separating a mixture of water and alcohol. The alcohol has a lower boiling point and escapes the liquid mixture faster than the water. As you study the diagram, start on the left and trace the process. Think about what is happening in each part of the diagram.

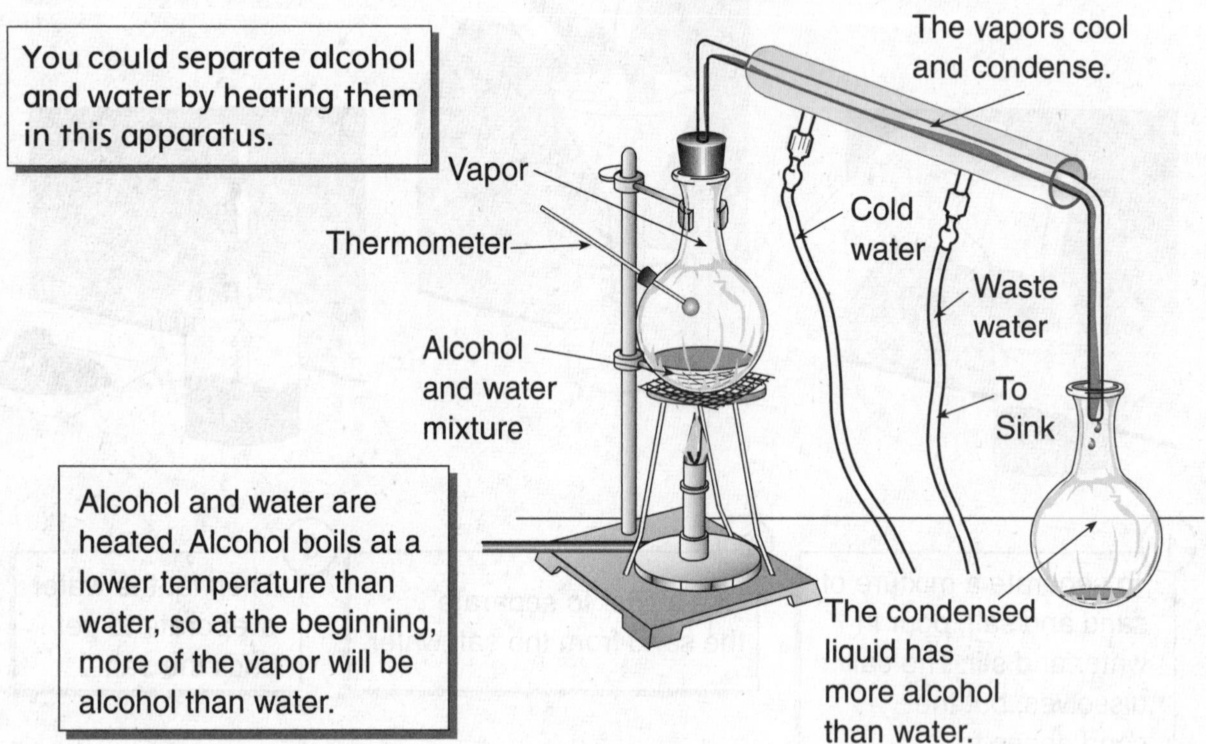

You could separate alcohol and water by heating them in this apparatus.
The vapors cool and condense.
Vapor
Thermometer
Cold water
Waste water
Alcohol and water mixture
To Sink
Alcohol and water are heated. Alcohol boils at a lower temperature than water, so at the beginning, more of the vapor will be alcohol than water.
The condensed liquid has more alcohol than water.

Answer these questions about the diagram above.

1. What is happening in the flask on the left?

____

2. Where do you think the gas particles in the tube above the flask came from? Why? ____

3. What is happening in the middle of the long tube?

____

4. What liquid would you expect to find in the flask on the right?

____

Name________________________ Date____________

Lesson Vocabulary
Lesson 4

Mixtures and Solutions

Match the correct letter with the description.

Vocabulary
a. mixture
b. solution
c. suspension
d. colloid
e. emulsion
f. aerosol
g. gel
h. foam
i. chemical
j. physical
k. heterogeneous

_____ **1.** a liquid spread through another liquid

_____ **2.** a mixture of substances that settle into layers in a fluid

_____ **3.** two or more parts blended together yet keeping their own properties and not turning into a new substance

_____ **4.** This kind of change results in a mixture.

_____ **5.** a solid spread through a liquid

_____ **6.** liquid drops or solid particles spread through a gas

_____ **7.** This forms when one substance is insoluble, or does not dissolve in a solvent.

_____ **8.** a mixture in which substances are completely blended so that the properties are the same throughout and the substances stay blended

_____ **9.** particles (or droplets) large enough to block out light spread through-out another substance

_____ **10.** This kind of combining results in a new substance.

_____ **11.** a gas spread through a liquid or solid

Use the terms above to describe the mixtures below. Then write if they are easy or hard to separate.

12. air ________________ ________________

13. a jar of coins ________________ ________________

14. a rock ________________ ________________

15. gelatin dessert ________________ ________________

16. whipped cream ________________ ________________

17. Give an example of how an aerosol is used, how a foam is used.

____

____

Name ______________________ Date ____________

Cloze Test
Lesson 4

Mixtures and Solutions

Vocabulary

solution	heterogeneous	emulsions	properties
suspensions	colloid	homogeneous	separate

Fill in the blanks.

A mixture is the physical combination of two or more substances. The substances in a mixture keep their own ________________ without forming new substances. In a ________________, the substances are blended completely. The mixture looks the same everywhere, or is ________________. In ________________ mixtures, the substances may be only partly blended. Examples of heterogeneous mixtures are ________________. They ________________ into layers when they are left alone. A third type of mixture, ________________, has properties in between those of solutions and heterogeneous mixtures. There are many types of colloids, such as ________________, aerosols, foams, and gels.

Name ______________________ Date ____________

Lesson Outline

Lesson 5

Chemical Changes

Fill in the blanks.  Reading Skill: **Cause and Effect** - questions 3, 7, 10, 12, 14

What Are Physical and Chemical Changes?

1. In a(n) ________________ change, matter changes in size, shape, or state without changing identity.
2. Chemical changes occur when ________________ link together in new ways.
3. A change in matter that produces a new compound with different properties from the original is a(n) ________________.
4. A chemical ________________ is a chemical change of original substances into one or more new substances.
5. A(n) ________________ is one of the original substances before a chemical change.
6. The new substances produced by a chemical reaction are called ________________.
7. In the reaction between baking soda and vinegar, the baking soda and vinegar are the reactants, and ________________, water, and sodium acetate are the products.

What Are the Signs of a Chemical Change?

8. Signs that a chemical change has occurred include:

   a. a(n) ________________ change,

   b. formation of a(n) ________________, and the formation of light and

   c. ________________.

9. When blueberry juice is mixed with a solution of baking soda, it turns a(n) __________________ color.

10. The __________________ that form when a solution containing dissolved baking soda and lemon juice are mixed indicates that a chemical change has taken place.

11. When you light a match, it gives off ______________ and ______________.

What Are Some Familiar Chemical Changes?

12. As a cake bakes, a chemical reaction turns the baking soda into sodium carbonate, steam, and __________________ gas, causing the cake to rise.

13. The chemical reaction involving baking soda is triggered by __________________.

14. The silver sulfide forms when silver reacts with sulfur or __________________ in foods or the air.

15. Iron oxide is commonly known as __________________.

Which Are Easier to Reverse—Chemical or Physical Changes?

16. Physical changes can sometimes be easily __________________.

17. In general chemical changes are __________________ to reverse.

Name________________________ Date____________

Interpret Illustrations

Lesson 5

What Are Physical and Chemical Changes?

This diagram shows what happens when baking soda and vinegar are mixed. Note how the atoms are combined into new substances.

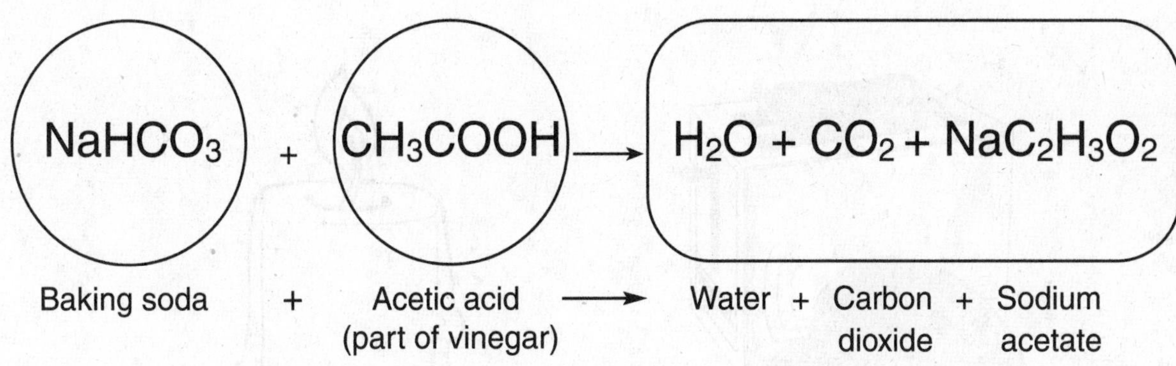

Answer these questions about the diagram above.

1. What elements make up

   Acetic acid? ______________________

   Baking soda? ______________________

2. What happens when the acetic acid and baking soda are mixed?

   ____

   ____

3. What kind of change, physical or chemical, is shown in the picture? How do you know?

   ____

   ____

4. What do you notice about the total number of atoms before and after the chemicals are mixed?

   ____

Name ______________________ Date ____________

Interpret Illustrations

Lesson 5

Which Are Easier to Reverse—Chemical or Physical Changes?

These drawings show changes that you may have seen at home. Think about how these changes happen. Also remember the differences between physical and chemical changes.

Answer these questions about the drawings above.

1. Is burning toast a physical or chemical change? How do you know?

____

____

2. What kind of physical change can you see when a candle burns? How can you tell it is a physical change?

____

____

3. What kind of chemical change can you see when a candle burns? How can you tell it is a chemical change?

____

____

Use with textbook page E76

Name______________________________ Date____________

Lesson Vocabulary

Lesson 5

Chemical Changes

Match the correct letter with the description.

Vocabulary
a. physical change
b. chemical change
c. chemical reaction
d. reactant
e. product

_____ 1. one of the new substances produced when a chemical reaction takes place

_____ 2. a change in size, shape, or state, without forming a new substance

_____ 3. a chemical change of original substance(s) into one or more new substances

_____ 4. one of the original substance(s) before a chemical reaction takes place

_____ 5. a change in matter that produces a new substance with different properties from the original

Identify the following as "chemical" or "physical."

6. a change in size not the substance ______________

7. produces heat ______________

8. a change in shape not the substance ______________

9. produces a gas not originally present ______________

10. change of state ______________

11. produces a new color from a new substance ______________

12. tarnish ______________

13. baking a cake ______________

14. burning a candle ______________

Name ______________________ Date ____________

Cloze Test
Lesson 5

Chemical Changes

Vocabulary

chemical reaction	products	rust	physical
state	reactants	chemical change	

Fill in the blanks.

In a ____________ change, matter may change its size, its shape, or its ____________ without also changing identity. When atoms link together to form new substances, a ____________ takes place. The ____________ are the original substances. New substances produced when a ____________ takes place are called the ____________. A common example of this kind of change is ____________ or iron oxide.

Name ______________________ Date ____________

Lesson Outline
Lesson 6

Acids and Bases

Fill in the blanks.  Reading Skill: **Cause and Effect** - questions 1, 2, 6, 7, 9, 12, 16, 18

What Are Acids and Bases?

1. A(n) ______________ tastes sour and turns blue litmus paper red.
2. Ammonia and baking soda, which are ______________, taste bitter and turn red litmus paper blue.
3. Acids give away hydrogen particles, or ______________ ions in a reaction.
4. Particles that are made up of one oxygen and one hydrogen atom linked together are called ______________ ions.
5. Water is neither acidic nor basic and is therefore ______________.

How Can You Tell if Something Is an Acid or a Base?

6. A(n) ______________ is a substance that changes color when it is mixed with an acid or a base.
7. A common vegetable indicator, ______________, is colorless in acid and turns pink in a base.

How Can Indicators Be Useful?

8. A test kit has paper strips that are soaked with an indicator to reveal how ______________ the soil is.
9. Aquarium owners or pond owners periodically have to check how acidic the ______________ is to make sure it is healthy for the fish and plants.

Name ______________________ Date ____________

Lesson Outline
Lesson 6

Are All Acids and Bases the Same?

10. The strength of an acid is its ____________.

11. The strength of a base is its ____________.

12. Acidity and alkalinity depend on how many hydronium and hydroxide ____________ there are in a solution.

13. Soren Sorenson developed the ____________ scale in 1909 to compare acidic and basic solutions.

14. A pH of ____________ means that there are a lot of hydronium ions in the solution, and the solution is very acidic.

15. A pH of ____________ means that there are a lot of hydroxide ions in a solution, and that the solution is very basic.

16. pH ____________ changes to a different color depending on the exact pH of a solution.

How Can Acids and Bases Be Used?

17. Many ____________, like tomatoes, grapefruit, lemons, and limes, get their flavor from acids.

18. A product that helps to increase the pH of your stomach back to its normal level is a(n) ____________.

What Are Acids and Bases?

The pictures below show models of a hydronium ion (H^+) and a hydroxide ion (OH^-). Acids give off hydronium ions and bases give off hydroxide ions. A substance which give off neither of these ions is neutral.

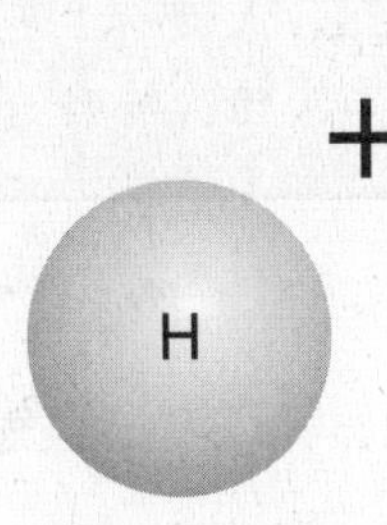

Hydronium ion

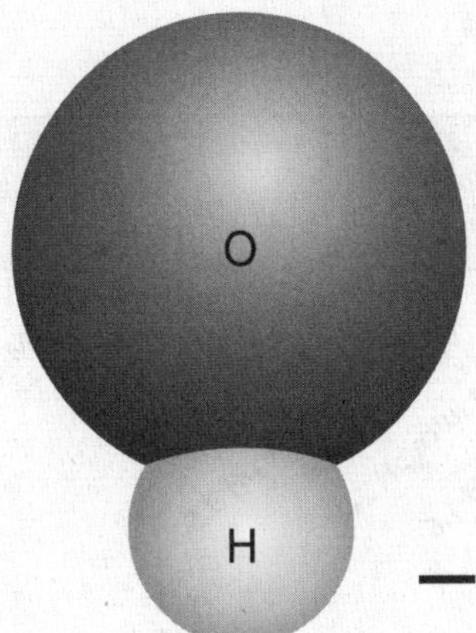

Hydroxide ion

Answer the following questions about the picture above.

1. Which ion contains oxygen?

____

____

2. If a solution gives off neither hydronium nor hydroxide ions, what does that indicate?

____

____

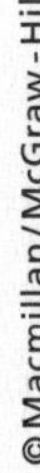

Name ______________________ Date ____________

Interpret Illustrations

Lesson 6

Are All Acids and Bases the Same?

The picture below shows a pH scale. The scale goes from 1, the most acidic, to 14, the most basic. pH 7 indicates a neutral substance on the scale.

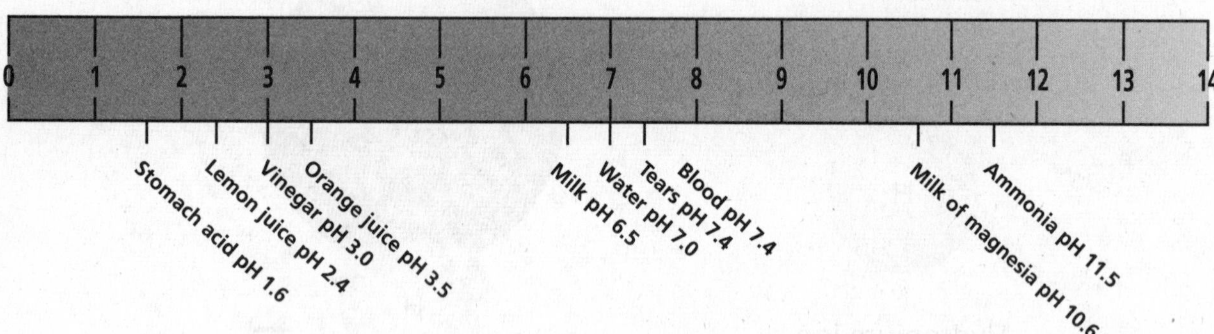

Answer the following questions about the picture above.

1. Which substance is more acidic—lemon juice or orange juice?

   ____

2. Which of the following would help to counteract stomach acids better—vinegar or milk?

   ____

3. Which solution is neutral?

   ____

Name ______________________ Date ____________

Lesson Outline
Lesson 6

Acids and Bases

Vocabulary

acidity

acid

pH

indicator

alkalinity

neutral

hydronium

base

Fill in the blanks.

1. Acidity and ________________ depend on how many hydronium and hydroxide ions are in a solution.
2. Something that tastes bitter and turns red litmus paper blue is a(n) ________________.
3. Hydroxide and ________________ ions combine to make water.
4. Litmus paper is an example of a(n) ________________ because its color changes when it is mixed with an acid or a base.
5. The strength of an acid is called its ________________.
6. A solution that is neither acidic nor basic, such as water, is a(n) ________________ substance.
7. The ________________ of a solution is a measure of the solution's acidity or alkalinity.
8. Something that gives off hydrogen particles or hydronium ions is a(n) ________________.

Answer each question.

9. Describe how red cabbage juice can be used as an indicator.

____

____

____

10. Who developed the pH scale and how is it used to describe acids and bases?

____

____

____

Name________________________________ Date______________

Cloze Test
Lesson 6

Acids and Bases

Vocabulary

neutral	base	strength	alkalinity
acid	indicator	hydronium	hydroxide

Fill in the blanks.

In a reaction, acids give off particles called ________________ ions. Bases give off ________________ ions that are made up of one oxygen and one hydrogen atom linked together. A substance that is ________________ does not release either hydronium or hydroxide ions. You can make a(n) ________________ neutral by mixing it with a base. A(n) ________________, such as litmus paper, changes color when it is mixed with an acid or a ________________. An indicator can be used to measure the ________________ of an acid or its acidity. Acidity and ________________, the strength of a base, depend on how many hydronium and hydroxide ions are in a solution.

Name________________________ Date____________

Lesson Outline

Lesson 7

Matter and Energy

Fill in the blanks.  Reading Skill: **Cause and Effect** - questions 3, 12

What Is Electrical Energy?

1. A battery provides ________________ energy.
2. The chemical reactions in a battery produce ________________, each of which carries energy.
3. When a battery is connected to a light bulb with wires, a(n) ________________ is formed.
4. When electrons in a circuit pass through a light bulb, they give up some of their energy to the wire in the light bulb, called the ________________.
5. The ________________ measures amps, telling us how many electrons flow each second.
6. The ________________ measures volts.

What Are Other Forms of Energy?

7. Energy is a measure of how much ________________ something can produce.
8. Work is using ________________ to move an object.
9. Energy is not a type of ________________.
10. Any object that is moving has ________________ energy.
11. Stored energy is referred to as ________________ energy.

Name ______________________ Date __________

Lesson Outline

Lesson 7

How Can We Describe Thermal Energy?

12. The transfer of thermal energy from one object to another is ______________.

13. Heat always flows from ______________ materials to cooler materials, never the other way.

How Does Heat Move?

14. In ______________, thermal energy flows through objects as their particles vibrate.

15. In ______________ hot parts of a material rise while cooler parts sink.

16. Heat is transferred through electromagnetic rays in ______________.

What Materials Conduct Heat Well?

17. Metals are the best ______________ of heat.

18. The molecules in ______________ are packed very closely together, and this makes transferring energy easier.

Name ______________________ Date ____________

Interpret Illustrations

Lesson 7

What Is Electrical Energy?

The picture below shows a flow of electrons from a battery through a light bulb's filament. Attaching the positive and negative terminals of the battery to the light bulb creates a closed circuit through which electrons can flow. The flow of electrons causes the filament to give off energy as light and heat.

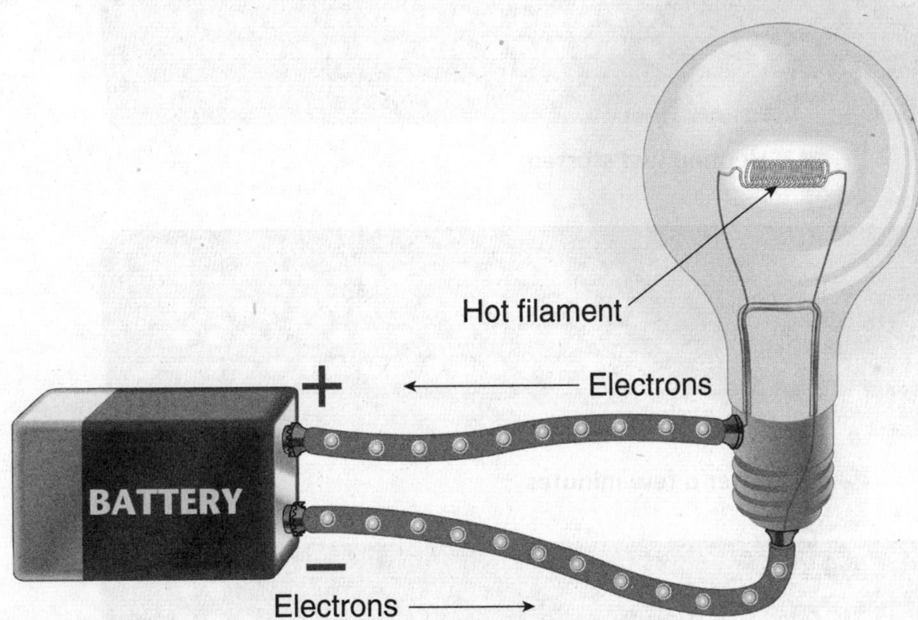

Answer the following questions about the picture above.

1. How do electrons flow through the circuit?

____

____

____

2. The filament gives off what kinds of energy?

____

____

3. Would the light bulb work if you attached both of the battery's terminals to one side of the light bulb? Why?

____

____

Name________________________ Date____________

Interpret Illustrations

Lesson 7

How Does Heat Move?

These three illustrations show different times during the heating process. In this case, a piece of metal is being heated. Start at the top and notice how each picture is different from the one below it.

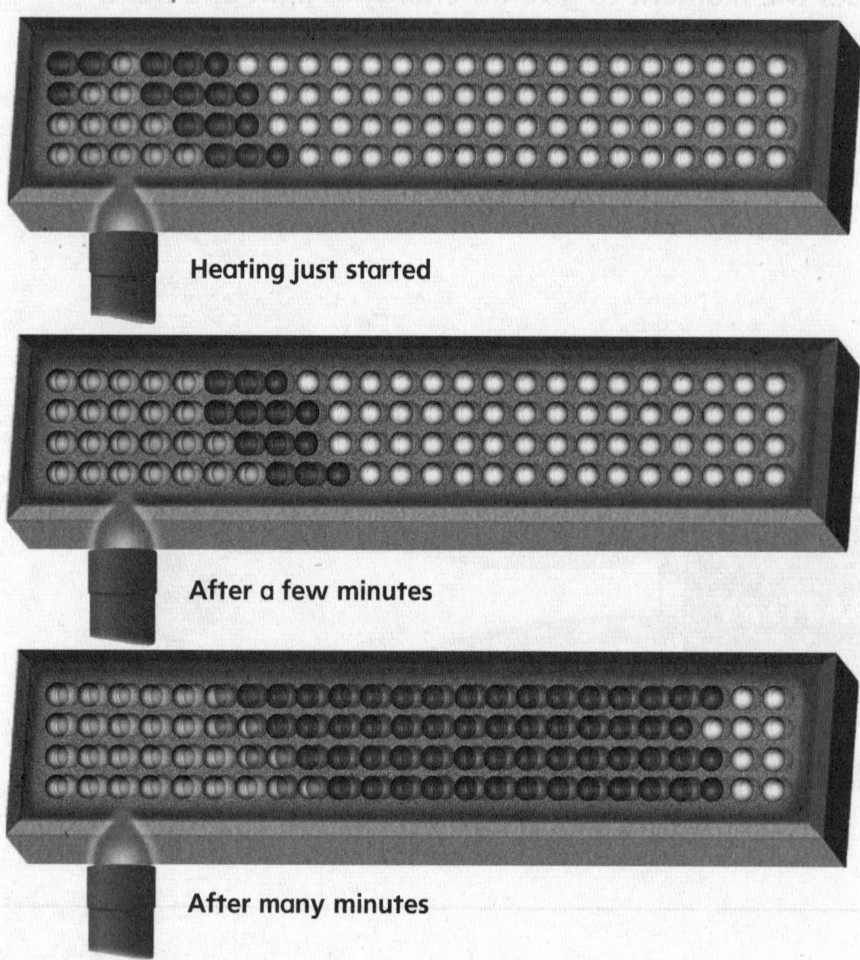

Answer these questions about the illustrations above.

1. Which metal particles are the first to react to the heat?

____

2. How do they react to the heat?

____

3. How does heat travel down the piece of metal?

____

____

Name ________________________ Date ____________

Lesson Outline
Lesson 7

Matter and Energy

Vocabulary

a. kinetic energy
b. potential energy
c. conduction
d. convection
e. radiation
f. filament
g. energy

Match the correct letter with the description.

_____ 1. the ability to move matter around

_____ 2. when hot parts of a material rise, while cooler parts sink

_____ 3. the energy of a moving object

_____ 4. a thin wire in a bulb

_____ 5. energy stored in an object or material

_____ 6. when heat is transferred through electromagnetic rays

_____ 7. movement of energy from a hot object that comes into contact with a cooler object; the material remains in place

Identify the form of energy and write one way you use that kind of energy.

8. carried by light energy ______________________

9. stored in links between atoms ______________________

10. sum of the kinetic and potential energy of a system

____

11. carried by electrons in circuits ______________________

12. heat produced by filament ______________________

Name ______________________ Date ____________

Cloze Test
Lesson 7

Matter and Energy

Vocabulary

potential energy	chemical energy	kinetic energy	work
force	matter	forms	

Fill in the blanks.

What is energy? Energy is not a type of ______________. It is a measure of how much ______________ something can produce. To scientists, work means using a ______________—a push or a pull—to move an object. There are many ______________ of energy. When something moves, its energy is called ______________. Another main form of energy is ______________. This energy can be described as stored energy. An example of this is ______________, which is the energy stored in links between atoms.

Name ______________________ Date __________

Chapter Vocabulary

Chapter 13

Forms of Matter and Energy

For each number, circle the letter of the best answer.

1. When a liquid turns blue litmus paper red, this indicates it is a(n)
   a. colloid. b. emulsion.
   c. solution. d. acid.

2. A mixture in which substances are completely blended so that the properties are the same throughout and the substances stay blended describes a(n)
   a. colloid. b. emulsion.
   c. solution. d. suspension.

3. When some substance tastes bitter and turns red litmus paper blue, it is a(n)
   a. colloid. b. base.
   c. solution. d. suspension.

4. A mixture with properties in between those of solutions and heterogeneous mixtures is called a(n)
   a. colloid. b. alloy.
   c. indicator. d. suspension.

5. Two or more substances blended together that keep their own properties describes a(n)
   a. compound b. base.
   c. mixture. d. molecule.

6. The strength of a base is called its
   a. alkalinity. b. emulsion.
   c. mixture. d. solution.

7. When thermal energy flows through objects that are touching, it is called
   a. radiation. b. conduction.
   c. convection. d. suspension.

8. A change in matter that produces a new substance with different properties from the original is a
   a. chemical change.
   b. kinetic change.
   c. physical change.
   d. product change.

9. When something moves, its energy is called
   a. kinetic.
   b. potential.
   c. physical.
   d. radiation.

10. This is one of the original substances before a chemical reaction takes place.
   a. colloid
   b. conductor
   c. product
   d. reactant

11. This is one of the new substances produced when a chemical reaction takes place.
   a. colloid
   b. conductor
   c. product
   d. reactant

12. Energy stored in an object or material is
   a. conduction energy.
   b. convection energy.
   c. kinetic energy.
   d. potential energy.

13. Movement of energy by the flow of matter from place to place is
   a. conduction.
   b. convection.
   c. emulsion.
   d. radiation.

14. Movement of energy that can travel through empty space is
   a. conduction.
   b. convection.
   c. emulsion.
   d. radiation.

Use with textbook pages E48–E101

Name ______________________ Date __________

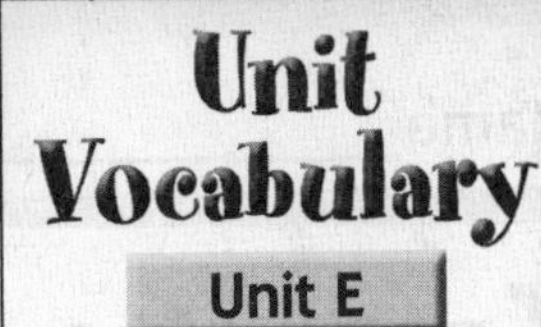

Choose-a-Word

Circle the word that best completes each sentence.

1. Vinegar, orange juice, and lemon juice are ______________.

   bases **acids** **neutrals**

2. Pure substances that can't be broken down are called ______________.

   compounds **hydrocarbons** **elements**

3. The original substances in chemical reactions are called the ______________.

   reactants **products** **molecules**

4. The ______________ of a solution tells how acidic or basic the solution is.

   potential **indicator** **pH**

5. An ______________ is the smallest unit of an element that still has the properties of that element.

   electron **atom** **emulsion**

6. A ______________ is formed by two or more elements and acts like a single substance.

   molecule **colloid** **compound**

7. If something is ______________, it prevents heat from passing through.

   instructed **insulated** **radiated**

8. The melting point of a substance is also known as the ______________ point.

   boiling **freezing** **evaporation**

9. A ______________ has properties in between those of solutions and heterogeneous mixtures.

   compound **suspension** **colloid**

Name ______________________________ Date ____________

Unit Vocabulary

Unit E

Correct-a-Word

Word Box				
potential	emulsion	mass	kinetic	volume
physical	boiling	radiation	gel	

One word in each sentence below is wrong. Cross it out and write the correct word above it. One has been done for you. Use the Word Box to check your spelling.

1. The amount of matter in an object is its ~~density.~~ mass
2. The energy of any moving object is kindling energy.
3. The vacuum describes how much space a sample of matter takes up.
4. A gem is a solid spread through a liquid.
5. The melting point is when a substance changes from liquid to gas.
6. In a physics change, matter changes in size, shape, or state without also changing identity.
7. One liquid is spread through another in an emotion.
8. Stored energy is pretentious energy.
9. The giving off of electromagnetic rays through space is radical.

Word Wheel

Start with the letter A. Write it on the first line below the puzzle. Now write down every third letter. (The next letter has a star next to it to help you.)

Continue around the wheel, writing down every third letter. After you go around three times, you'll have an important message.

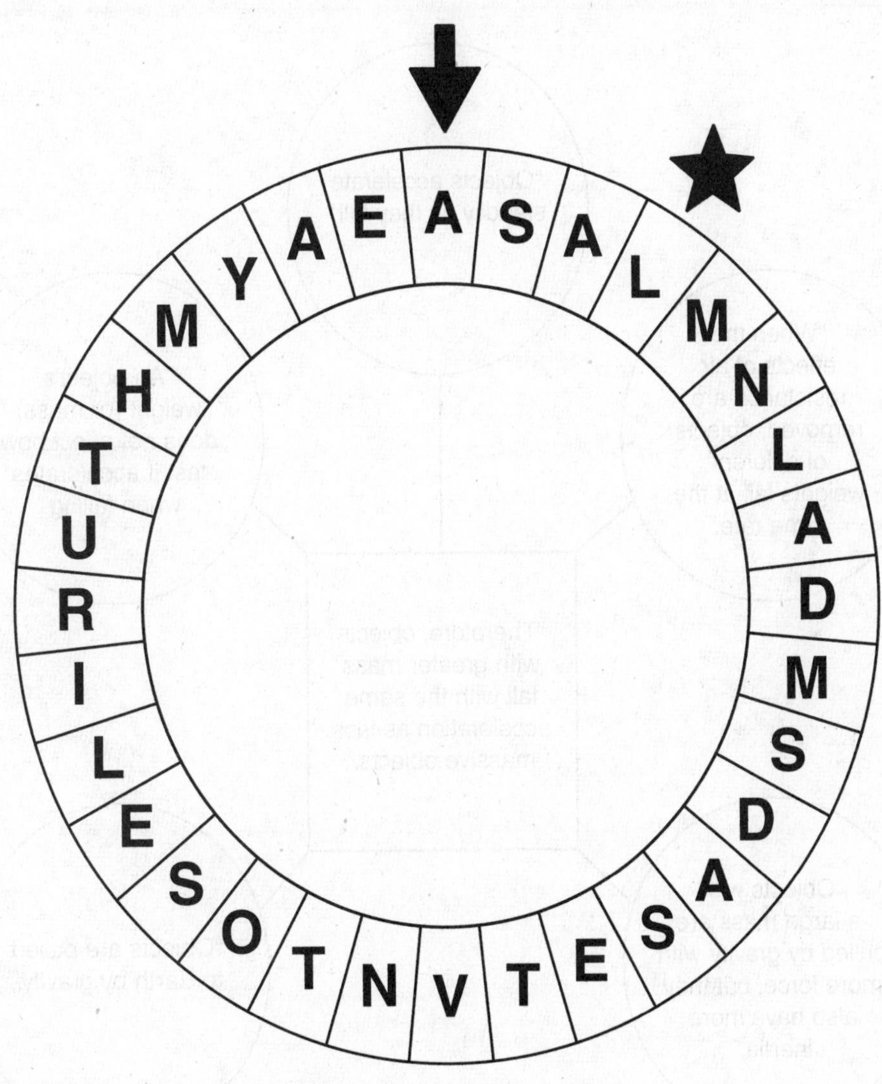

___ ___ ___   ___ ___ ___ ___ ___ ___   ___ ___ ___

___ ___ ___ ___,   ___ ___ ___ ___ ___ ___,

___ ___ ___   ___ ___ ___ ___ ___ ___ ___!

Name______________________________ Date______________

Chapter Graphic Organizer

Chapter 14

Newton's Laws of Motion

Make an idea web like the one shown below on a separate piece of paper. List facts from your book in each of the outer web circles. Then write your conclusion in the center square. See if your conclusion is correct by checking it against each of the facts that you listed. Each fact should support your conclusion. Look at the example below.

"Objects accelerate steadily as they fall."

"When the effects of air resistance are removed, objects of different weights fall at the same rate."

"An object's weight (or mass) does not affect how fast it accelerates when falling."

"Therefore, objects with greater mass fall with the same acceleration as less massive objects."

"Objects with a large mass are pulled by gravity with more force, but they also have more inertia."

"Objects are pulled to Earth by gravity."

Name ______________________ Date ____________

Draw Conclusions

Conclusions should be based on facts and evidence, not on guesswork. It can help to number the known facts and separately list them. After proposing your conclusion, work backwards and see if each of the facts supports your conclusion.

Look at the paragraph below about acceleration. List the facts that lead to the conclusion below.

What if you use a rubber band to launch a toy car along the floor? The rubber band will apply a force to the car, and the force will cause the car to speed up. Once the rear of the car passes the starting line, however, the rubber band no longer is applying force. At this point the car will begin coasting until friction brings it to a stop. The farther the car travels before stopping, the faster it must have been going at the start. When you add extra rubber bands you are applying more force to the car. As the force increases, the distance the car travels also increases.

Conclusion: The car reaches its greatest starting speed when the force applied to it is greatest.

Name ______________________ Date ____________

Chapter Reading Skill
Chapter 14

More Drawing Conclusions

Conclusions are based on facts and evidence. Read the paragraph below.

After reviewing the facts in the paragraph, underline the correct conclusion from the following list.

There is a story that says that a falling apple may have set Isaac Newton to thinking about gravity. In the late 1660's there was a plague (very bad illness that spread very easily) in Cambridge, England where Newton had gone to college. To avoid the plague, he went home to the countryside. The legend says Newton was sitting under an apple tree one day, when an apple hit him on the head. The legend may or may not be true. However, an idea did hit Newton. That idea was that the force that pulls an apple to the ground is the same force that keeps the moon in its orbit around Earth.

Conclusion

1. Isaac Newton got the plague and that caused him to discover gravity.
2. The apple falling on Newton's head told him about the concept of gravity.
3. The legend about an apple falling on Newton's head is not true but he did come up with an idea about forces.
4. Newton came up with the idea that the force that pulls an apple to the ground is the same force that keeps the moon in its orbit around Earth.

Name______________________________ Date______________

Lesson Outline
Lesson 1

Newton's First Law

Fill in the blanks.  Reading Skill: **Draw Conclusions** - questions 6, 9, 15

What Does It Take to Make an Object Move?

1. A pull or push that acts on an object is called a ______________.
2. Forces are needed to set objects in ______________.
3. The ______________ of an object tends to make the object resist being set into motion.
4. The tendency of an object to resist a change in its state of motion is called the object's ______________.
5. Galileo imagined rolling a ball down a fixed ______________ and then back up ramps of varying steepness.
6. Without a force to stop or slow a coasting object, the object will coast forever in a ______________ line.

Is Force Needed to Maintain Motion?

7. ______________ opposes the motion of one object moving past another.
8. ______________ published a complete description of the concept of inertia.
9. Objects at rest remain at rest and objects traveling at a steady rate in a straight line continue that way until a ______________ acts on them.
10. ______________ first law of motion tells us that if an object is sitting at rest, it will continue to be at rest until a force is applied to it.

Name ______________________ Date ____________

Lesson Outline

Lesson 1

Where Is It?

11. You are moving when you are changing ______________.

12. Position is the ______________ of an object.

13. The ______________ of a moving object is how fast its position is changing with time at any moment.

What Is Velocity?

14. The speed of a moving object taken together with its direction of travel gives the ______________ of the object.

15. Two objects can have the same speed but different velocities if they are traveling in different ______________.

What Is Acceleration?

16. As long as an object travels in a straight line at a steady speed, its velocity is ______________.

17. A change in velocity is called ______________.

18. ______________ occurs when a force causes the speed of an object to decrease.

What Keeps Things Moving In a Circle?

19. ______________ is the force that keeps Earth moving in a circular path about the Sun.

Name ______________________ Date ____________

Interpret Illustrations
Lesson 1

What Does It Take to Make an Object Move?

Galileo reasoned, a ball rolling down one ramp and up another would roll to the same height on any ramp. He tested his hypothesis in an experiment to determine how inertia affects the motion of a ball rolling down a fixed ramp and then up ramps of varying steepness as shown in the picture below.

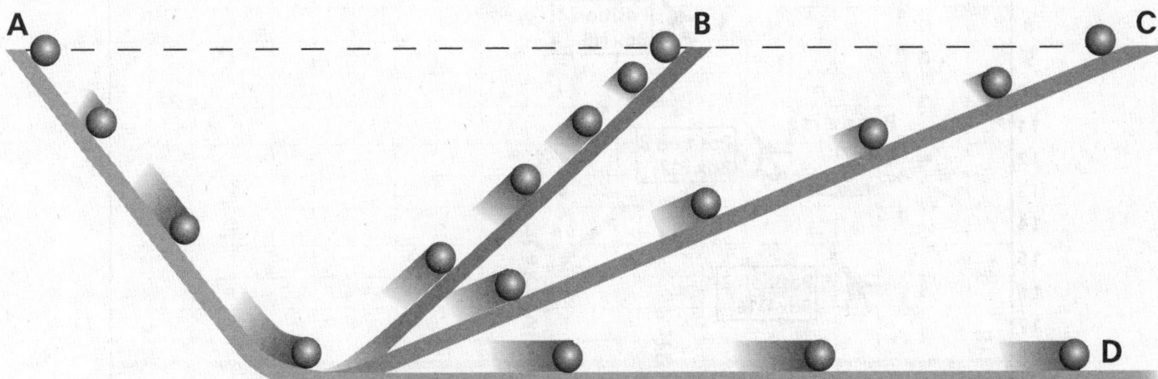

Answer the following questions about the picture above.

1. What might be acting on the ball to make it stop at points B and C?

____

2. Could you assemble a ramp that allowed a ball that started at point A to rise to a height that would be higher than the line drawn through points A and B?

____

____

3. If the ball traveled from point A up ramp B and stopped at point B, would it remain there indefinitely?

____

____

Name________________________ Date____________

Interpret Illustrations

Lesson 1

Where Is It?

The diagram below shows the shadow of an airliner moving over a town at a steady speed. One second passes between each position.

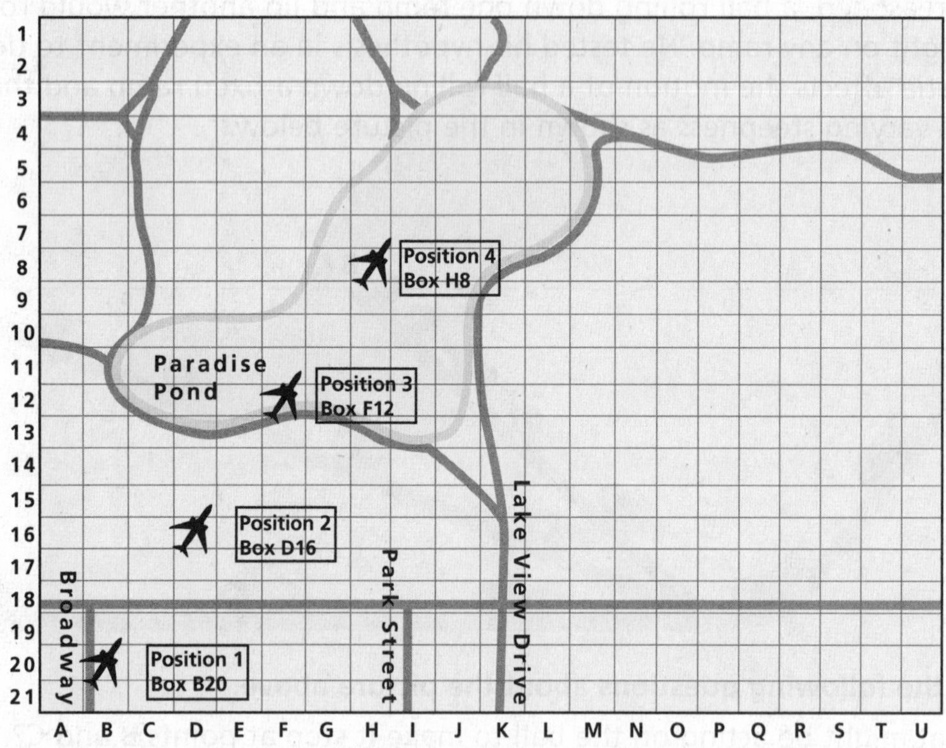

Answer the following questions about the picture above.

1. If the plane continues to travel in the same direction and reaches **I7** on the grid one second after passing through **H8,** has its speed changed?

____

____

2. How would the speed of the plane have changed if the plane reaches **I7** on the grid two seconds after passing through **H8?**

____

____

3. How would the speed of the plane have changed if the plane reaches **L1** on the grid in less than a second after passing through **H8?**

____

____

Name ______________________________ Date ______________

Lesson Vocabulary

Lesson 1

Newton's First Law

Vocabulary

deceleration
velocity
acceleration
speed
inertia
friction
position
force

Fill in the blanks.

1. A force called ____________________ opposes the motion of one object moving past another.
2. ____________________ is the location of an object.
3. A pull or push that acts on an object is called a ____________________.
4. A change in velocity is called ____________________.
5. The ____________________ of an object is how fast its position is changing with time at any moment.
6. The tendency of an object to resist a change in its state of motion is called the object's ____________________.
7. ____________________ occurs when a force causes the speed of an object to decrease.
8. The speed of a moving object taken together with its direction of travel gives the ____________________ of an object.

Answer each question.

9. What are two ways that an object's velocity can change?

____

____

10. Describe two forces that can affect an object's velocity.

____

____

Name________________________________ Date______________

Cloze Test
Lesson 1

Newton's First Law

Vocabulary

inertia	deceleration	gravity	straight
velocity	law	direction	acceleration

Fill in the blanks.

As long as an object travels in a straight line at a steady speed, its ____________________ is constant. Newton's first ____________________ tells us that an object's velocity will remain constant unless a force is applied to it. A force may change an object's speed, or its ____________________ of travel, or both. A change in an object's velocity is called ____________________. Newton realized that applying a force to an object would overcome its ____________________ and change its velocity, causing it to accelerate. A special case of acceleration, called ____________________, occurs when a force causes the speed of an object to decrease. A force called ____________________ pulls on Earth and keeps Earth moving in a circular path about the Sun. Without the force of gravity, Earth would fly off in a ____________________ line into deep space.

Name ______________________ Date ____________

Lesson Outline
Lesson 2

Newton's Second and Third Laws

Fill in the blanks.  Reading Skill: **Draw Conclusions** - questions 3, 5, 15, 16

What Affects Acceleration?

1. If a force is applied to an object, the object's ____________________ will change.
2. If the ____________________ applied to an object is multiplied by a certain amount, the acceleration will change by the same amount.
3. Doubling the mass of an object results in ____________________ the acceleration.
4. Increasing the force increases the acceleration, while increasing the ____________________ decreases the acceleration.
5. The relationship between acceleration and mass is ____________________ —when one goes up, the other goes down.

How Is Acceleration Calculated?

6. A force of 1 ____________________ makes the speed of a 1-kg mass change by 1 meter per second each second.
7. We can write a(n) ____________________ to show how change in speed is related to force and mass.

What Are Balanced and Unbalanced Forces?

8. When all of the forces on an object cancel one another out, the forces are said to be ____________________ forces.
9. In cases where a certain force is either only partially canceled or not canceled at all by other forces, the force is said to be an ____________________ force.

Name ______________________________ Date ____________

Lesson Outline

Lesson 2

Fill in the blanks.

Where Does the Force Come From?

10. If an object is accelerating it must have an ____________ force acting on it.

How Do Forces Act Between Objects?

11. When one object applies a force to a second object, that force is called the ____________.

12. The force that a second object returns to the first object is called the ____________.

13. Newton realized that while the action and reaction act in opposite directions, they have the same ____________.

14. Newton's ____________ law of motion states that for every action, there is an equal but opposite reaction.

How Do Forces Affect Us?

15. The water coming from a hose is under very high pressure, which applies a large ____________ back on the hose.

16. In a row boat, you push the water with oars and the water ____________ back on the oars to move the boat forward.

What Is a Simple Machine?

17. A simple machine makes ____________ easier.

18. The six types of simple machines are levers, wheels and axles, ____________, ____________, ____________, and ____________.

What Is an Inclined Plane?

19. The slanted surface of an inclined plane is the ____________.

20. The vertical end of an inclined plane is the ____________.

Name ______________________ Date __________

Interpret Illustrations

Lesson 2

Where Does the Force Come From?

The pictures below show the forces at work in a closed and an open balloon. The balloon can be used to illustrate Newton's second and third laws about balanced and unbalanced forces.

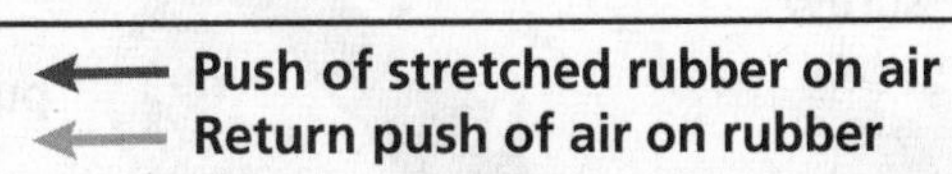

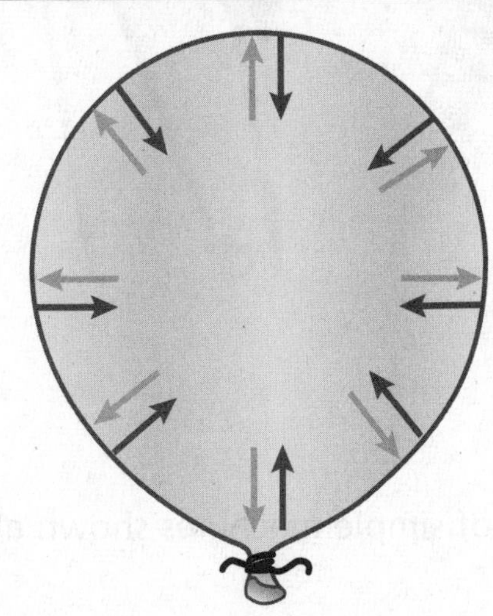

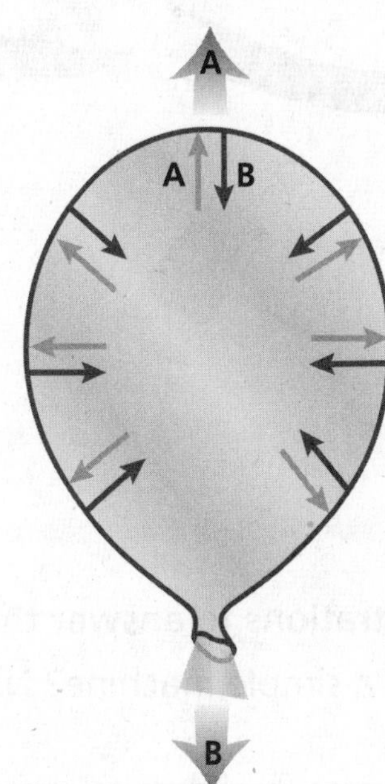

Answer the following questions about the picture above.

1. How are the forces inside of the closed balloon balanced?

____

____

2. Opening the balloon creates an unbalanced force that results in what kind of movement?

____

____

3. Which balloon can be used as a balloon rocket? Explain.

____

____

Name ______________________ Date ____________

Interpret Illustrations

Lesson 2

What is a Simple Machine?

The illustrations below show examples of four different types of simple machines.

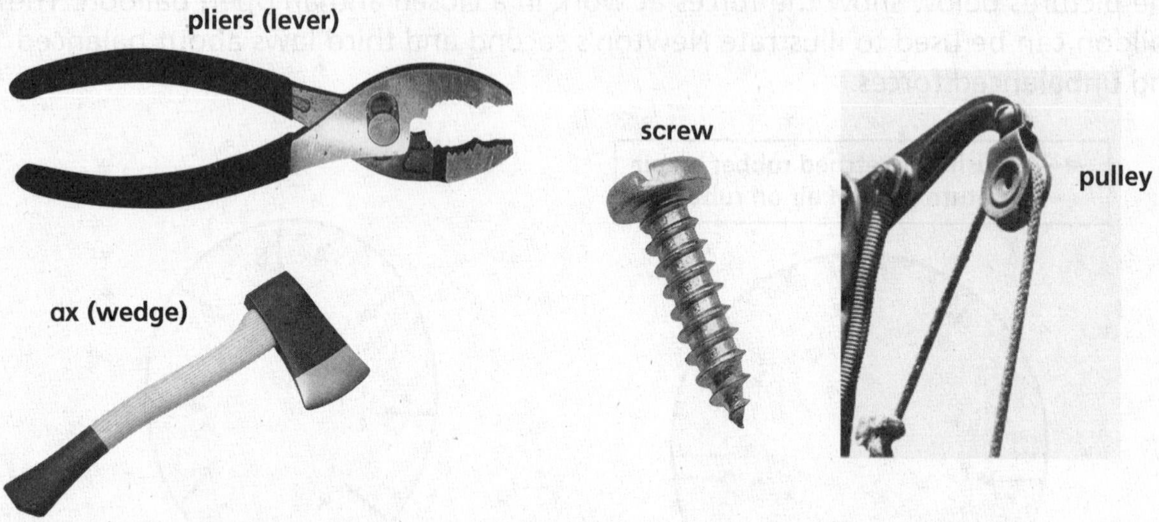

Use the illustrations to answer the questions.

1. What is a simple machine? Name the types of simple machines shown above.

____

____

2. Describe the direction of the force applied to a screw. What is the motion of the screw as force is applied?

____

____

3. How does pulley make work easier?

____

____

4. Is a screw a wedge? Explain your answer.

____

____

Name ______________________ Date __________

Lesson Vocabulary
Lesson 2

Newton's Second and Third Laws

Vocabulary

- lever
- balanced forces
- fulcrum
- simple machine
- reaction
- work
- unbalanced force
- effort arm
- resistance arm
- action

Fill in the blanks.

1. Using force to move an object through a distance is ______________.
2. A pulley is a(n) ______________.
3. In cases where a certain force is either partially canceled out or not cancelled out at all by other forces, the force is said to be a(n) ______________.
4. A ______________ has a rigid bar that rests on a ______________.
5. The force that a second object returns to the first object is called the ______________.
6. When all of the forces on an object cancel one another out, the forces are said to be ______________.
7. When one object applies a force to a second, we call this force the ______________.
8. With a lever, you apply force to the ______________ and the ______________ produces a force to lift the load.

Answer each question.

9. Explain how you might experience a change of acceleration in a car.

____

____

____

Name______________________________ Date______________

Cloze Test
Lesson 2

Newton's Second and Third Laws

Vocabulary

equal	reaction	action	third
force	strength	balanced forces	unbalanced force

Fill in the blanks.

A pull or push that acts on an object is called a(n) ____________________.

When all of the forces on an object cancel one another out, the forces are said to be ____________________. In cases where a certain force is either only partially cancelled or not cancelled at all by other forces, the force is said to be a(n) ____________________.

When one object applies a force to a second, we call this force the ____________________. The force the second object returns to the first is called the ____________________. Sir Isaac Newton realized that while the action and reaction act in opposite directions, they have the same ____________________. These ideas are summarized in Newton's ____________________ law of motion: For every action, there is an ____________________ but opposite reaction.

Name________________________ Date__________

Lesson Outline
Lesson 3

Newton's Law of Gravitation

Fill in the blanks. Reading Skill: **Draw Conclusions** - questions 1, 2

Why Would Air Make a Difference?

1. Air offers ________________ to the motion of objects through it.

What Makes Objects Fall at the Same Rate?

2. Galileo concluded that objects accelerate steadily as they fall and that an object's weight, or ________________ does not affect how fast it accelerates when falling.

3. An object is pulled to Earth by ________________, an attraction between the mass of Earth and the mass of an object.

4. Objects with a large mass are pulled on by gravity with more force, but they also have more ________.

What Is the Acceleration of Falling Objects?

5. The force of gravity on any object is called its ________________.

6. An object's weight in newtons can be found by multiplying its mass in kg by ________________.

7. The ________________ of any object falling to the ground increases by 9.8 meters per second each second.

8. A combination of the Moon's inertia and the force of gravity between Earth and the Moon keeps the Moon ________________ Earth.

How Can Gravity Be Universal?

9. Newton decided that as mass ____________________, the force of gravity also increases.

10. Newton's law of universal ____________________ states that the force of gravity between two objects increases with the mass of the objects and decreases with the distance between them squared.

11. For light objects the force of gravity is ____________________.

12. For ____________________ objects like moons, planets, and stars, the masses are so large that the force of gravity becomes very large.

When Is Added Weight Helpful?

13. We are accustomed to gravity giving things on Earth weight, including our own ____________________.

14. In cycling, the weight of the rider and bicycle presses the tires against the ground causing friction, which gives the tires ____________________ and drives the rider forward.

Name ______________________ Date ____________

What Makes Objects Fall at the Same Rate?

The picture below shows how gravity affects objects of different masses. According to Newton's second law, force equals mass multiplied by acceleration or F = ma. The acceleration of gravity changes slightly depending on how far away you are from Earth's center. However, since it varies only slightly, the acceleration of Earth's gravity can be rounded to equal 9.8 meters per second each second. Use this number to represent the acceleration in each of the following questions.

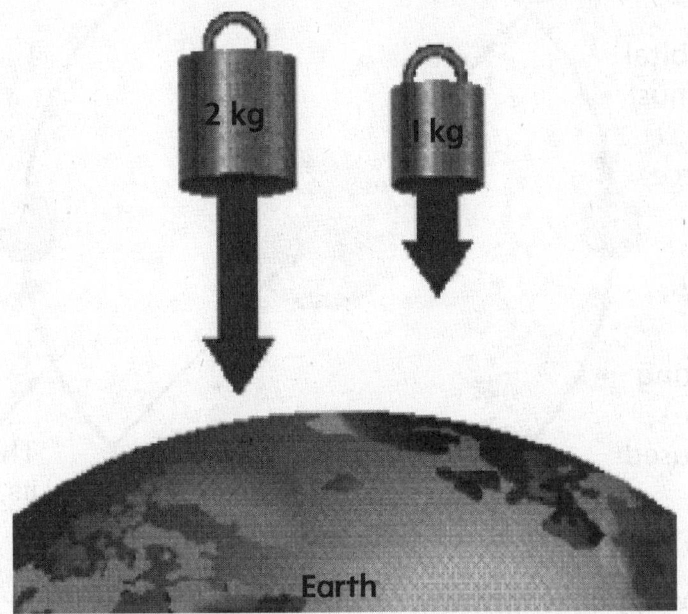

Answer the following questions about the picture above.

1. If the 1 kg object is being pulled by gravity by a force of 9.8 newtons, how many newtons of force does gravity exert on the 2 kg object?

____

____

2. If acceleration equals force divided by mass, what is the acceleration of each object if the force of gravity on the 1 kg object is 9.8 newtons and the force of gravity on the 2 kg object is 19.6 newtons? Note: acceleration is expressed as meters per second each second.

____

3. How many newtons of force would gravity exert on a 3 kg object?

____

Name________________________ Date____________

Interpret Illustrations

Lesson 3

What Is the Acceleration of Falling Objects?

The picture below shows the moon's orbit around Earth. The force of Earth's gravity keeps the moon accelerating towards Earth. Otherwise, the moon would fly off into space.

If there were no force pulling the Moon inward, it would fly off into space along this line.

To stay on the orbital path, the Moon must accelerate in this direction. The force pulling it inward causes the acceleration. Therefore, the Moon is accelerating toward Earth. The acceleration is caused by gravity.

The Moon's speed in its orbit is 1,030 m/scd

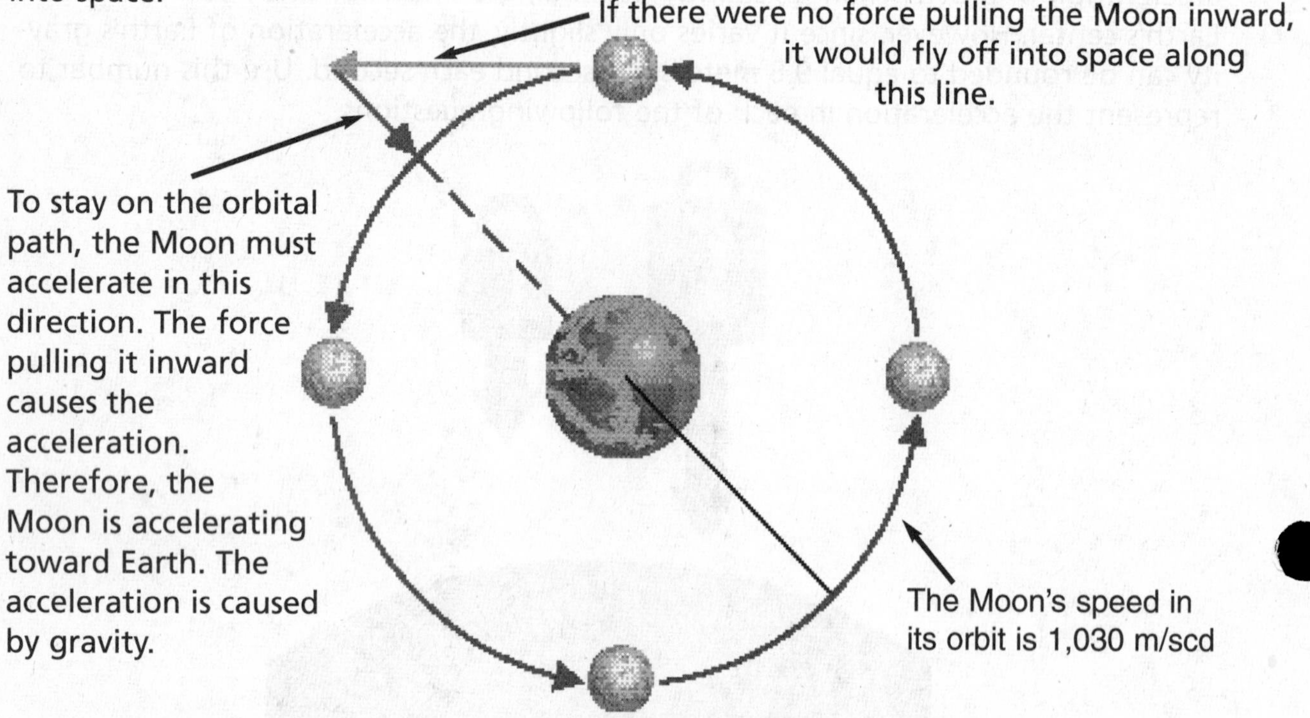

Answer the following questions about the picture above.

1. The moon moves at a constant speed of what around its orbit?

____

____

2. Is the moon's acceleration expressed as a change in speed or a change in direction?

____

____

3. If the moon was suddenly released from Earth's gravity what would happen to it? Would it still be experiencing acceleration?

____

____

Use with textbook page F37

Name ____________________ Date __________

Lesson Vocabulary
Lesson 3

Newton's Law of Gravitation

Fill in the blanks.

Vocabulary
gravity
inertia
resistance
universal
weight
distance
9.8

1. The force of gravity on any object is called the object's ______________.
2. The force of gravity between two objects increases with the mass of the objects and decreases with the ______________ between them squared.
3. If we ignore air resistance, all objects accelerate to the ground at ______________ meters per second each second.
4. Scientists have learned that when the effects of air ______________ are removed, objects of different weights fall at the same rate.
5. Newton's law of gravity is ______________ because it applies to any objects, not just moons, planets, and stars.
6. An object is pulled to Earth by ______________, an attraction between the mass of Earth and the mass of the object.
7. A combination of the Moon's ______________ and the force of gravity between Earth and Moon keeps the Moon orbiting Earth.

Answer each question.

8. Explain how added weight might help to keep a car on the road from spinning off a sharp turn. ______________

____

____

9. How did Galileo test his theories about objects of different masses falling at the same rate? ______________

____

____

Name________________________________ Date______________

Cloze Test
Lesson 3

Newton's Law of Gravitation

Vocabulary

weights	ramps	accelerate
Galileo	resistance	mass
gravity	Aristotle	inertia

Fill in the blanks.

____________________ believed that heavy things fall faster than lighter things. Later, ____________________ reasoned that objects fall at the same rate, ignoring air resistance. To test his ideas about falling objects, Galileo carried out experiments that involved rolling marbles down ____________________. He also talked about dropping two objects with different ____________________ off a tall tower to show that they would hit the ground at the same time. Galileo concluded that objects ____________________ steadily as they fall and that an object's weight or ____________________ does not affect how fast it accelerates when falling.

We know today that Galileo was right. An object is pulled to Earth by ____________________, an attraction between the mass of the Earth and the mass of the object. Objects with a large mass are pulled on by gravity with more force, but they also have more ____________________. This extra ____________________ to motion exactly offsets the greater pull of gravity on them. Therefore, objects with greater mass fall with the same acceleration as less massive objects.

Newton's Laws of Motion

Circle the letter of the best answer.

1. A force that opposes the motion of one object moving past another is
   a. gravity. b. friction.
   c. acceleration. d. unbalanced.

2. A pull or push that acts on an object is called a(n)
   a. acceleration. b. speed.
   c. force. d. inertia.

3. A change in velocity is called
   a. force. b. gravity.
   c. friction. d. acceleration.

4. How fast an object is changing its position with time at any moment is its
   a. speed. b. velocity.
   c. force. d. acceleration.

5. The tendency of an object to resist a change in its state of motion is called the object's
   a. speed. b. inertia.
   c. velocity. d. friction.

6. Velocity is
   a. how fast an object is changing its position with time at any moment.
   b. the acceleration of an object towards Earth.
   c. the rate at which friction slows the speed of an object.
   d. the speed of a moving object taken together with its direction of travel.

7. An attraction between the mass of Earth and the mass of the object is called
   a. mass. b. speed.
   c. gravity. d. velocity.

Name_______________________________ Date_______________

Chapter Vocabulary

Chapter 14

Circle the letter of the best answer.

8. An unbalanced force is
   a. a force that pushes in a particular direction.
   b. the rate of acceleration multiplied by the mass of an object.
   c. a force either partially canceled out or not cancelled out at all by other forces.
   d. when all forces in a system equally push on each other.

9. The force of gravity between two objects increases as their masses
   a. increase. b. decrease.
   c. decelerate. d. accelerate.

10. The force that a second object returns to the first object is called the
    a. inertia. b. action.
    c. unbalanced force. d. reaction.

11. When the forces on an object cancel each other out, the forces are said to be
    a. unbalanced. b. balanced.
    c. a reaction. d. massive.

12. The force of gravity on any object is called the object's
    a. inertia. b. velocity.
    c. reaction force. d. weight.

13. The force that prevents different objects from falling at the same rate is called
    a. inertia. b. gravity.
    c. air resistance. d. acceleration.

14. A simple machine that has a rigid bar and fulcrum is a(n)
    a. lever. b. bicycle.
    c. pulley. d. ramp.

Name______________________________ Date__________

Chapter Summary

1. What is the name of the chapter you just finished reading?

2. What are four vocabulary words you learned in the chapter? Write a definition for each.

3. What are two main ideas that you learned in this chapter?

Name_________________________ Date____________

Chapter Graphic Organizer
Chapter 15

Sound Energy

You can use a cause-and-effect diagram to show how one event causes a second event. Sometimes the second event can cause a third event, the third event can cause a fourth event, and so on. Complete the cause-and-effect diagram for **Sound Waves** shown below. The first step has already been filled in for you. Notice that the effect of the first step is also the cause of the second step. In the same way, the cause of the third step will be the same as the effect of the second step, and so on.

Sound Waves

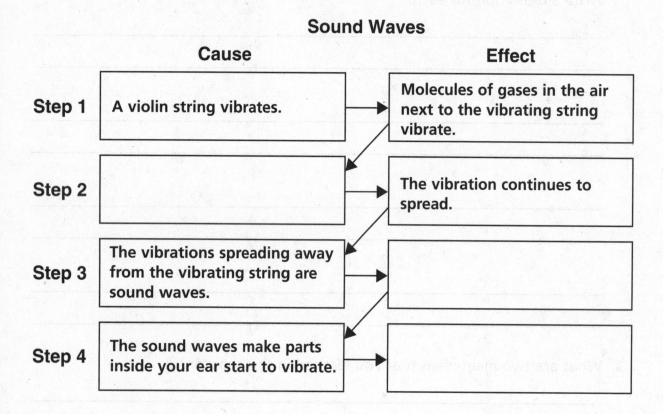

	Cause		Effect
Step 1	A violin string vibrates.	→	Molecules of gases in the air next to the vibrating string vibrate.
Step 2		→	The vibration continues to spread.
Step 3	The vibrations spreading away from the vibrating string are sound waves.	→	
Step 4	The sound waves make parts inside your ear start to vibrate.	→	

Name______________________________ Date____________

Chapter Reading Skill
Chapter 15

Cause and Effect

Things don't just happen. Living things and forces make them happen. Whatever or whoever makes something happen is the **cause;** what happens is the **effect.**

Here's an example. The frequency of a very high note sung by a singer can actually shatter glass. Vibrating sound waves are the cause, and a broken glass is the effect.

As you read books, look for clues to what makes things happen. Watch for clue words like because and so that point out a cause and its effect. For example: *I missed the bus* ***because*** *I overslept* or *I overslept,* ***so*** *I missed the bus.*

Read each statement below. Then circle the cause and underline the effect.

1. My baby brother cried because he fell down.
2. The glass shattered when the singer hit a high note.
3. Because of the snowstorm, we had to cancel the game.
4. I had no clean shirts, so I washed the dirty ones!
5. Because I was sick last week, I missed the big math test.
6. I broke my glasses, so I can't see the assignment written on the board!
7. He can't cross the street when the DON'T WALK sign is on.
8. I have a lot of allergies, so I can't have a kitten.
9. I tripped over the toys you left on the floor!
10. I don't have any money, so I can't buy that CD.

Now make up two cause-and-effect statements. Ask a classmate to identify each cause and effect.

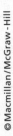

____

____

____

Name ______________________ Date __________

Chapter Reading Skill

Chapter 15

If . . . then

Some special cause-and-effect statements are called **If . . . then** statements. They state that if one thing happens, then another thing will happen.

Most **If . . . then** statements are true. However, **If** the first thing isn't true, **then** the second thing could never happen. Here's an example: ***If** this month is April, **then** next month will be May.* If it is really April, then the second part is true. However, if today is December 3, tomorrow can't possibly be a day in May!

Read the If . . . then statements below. If the first thing can make the second thing happen, write *T* for *true*. If it can't, write *NT* for *not true*.

______ **1.** If Marcy takes ballet lessons, then she must be a very good dancer.

______ **2.** If I do well on the test, then I'll get a good grade.

______ **3.** If my brother is 15 years old, then next year he'll be 16.

______ **4.** If it's very hot today, then it'll be cool tomorrow.

______ **5.** If a wheel is round, then it must be a circle.

______ **6.** If I have a library card, then I can sign out books and take them home.

______ **7.** If blue is my favorite color, then it must be your favorite color, too.

______ **8.** If I'm asleep in bed, then I must be sick.

______ **9.** If you get your hair cut, then your hair will be shorter.

______ **10.** If sound waves travel, then they have to bounce off objects in their paths.

Now write two of your own If . . . then statements. Read them to your classmates. Let them decide if it's T or NT.

____

____

____

____

Sound Waves

Fill in the blanks.  Reading Skill: **Cause and Effect** - questions 1, 5, 9, 11

What Makes Sound?

1. Sounds can be produced by a back and forth motion called a(n) ______________.
2. Many vibrations that make sound are too ______________ to see.
3. When you pluck a guitar string, you provide the ______________ necessary for this vibration.
4. The instruments in each section of the orchestra have their own characteristic ways of producing sounds, because in each section different ______________ vibrate.
5. What vibrates to produce sounds in the following instruments?
   a. drums, cymbals ______________
   b. trumpet, tuba ______________
   c. saxophone, clarinet ______________
6. Sound is a vibration that travels through ______________.
7. Matter can be in the form of a(n) ______________, a(n) ______________, or a(n) ______________.
8. Some types of matter are made of pieces too small to be seen; these are called ______________.
9. A vibrating string causes molecules of gas in the ______________ next to it to vibrate, too.
10. A vibration that spreads away from a vibrating object is called a(n) ______________.

Name ____________________________ Date ____________

Lesson Outline

Lesson 4

Fill in the blanks.

What Else Can Sound Go Through?

11. Inside the ear, sound waves cause the parts inside to ________________.

12. Sound waves can travel through ________________, such as air, and also through ________________ and ________________.

Name ______________________ Date ____________

Interpret Illustrations

Lesson 4

What Makes Sound?

The chart below shows how various musical instruments create sound waves. Each instrument can produce sound in different ways. The chart shows which part of the instrument vibrates. Each category of instruments has a different part that vibrates.

How Sound Is Produced

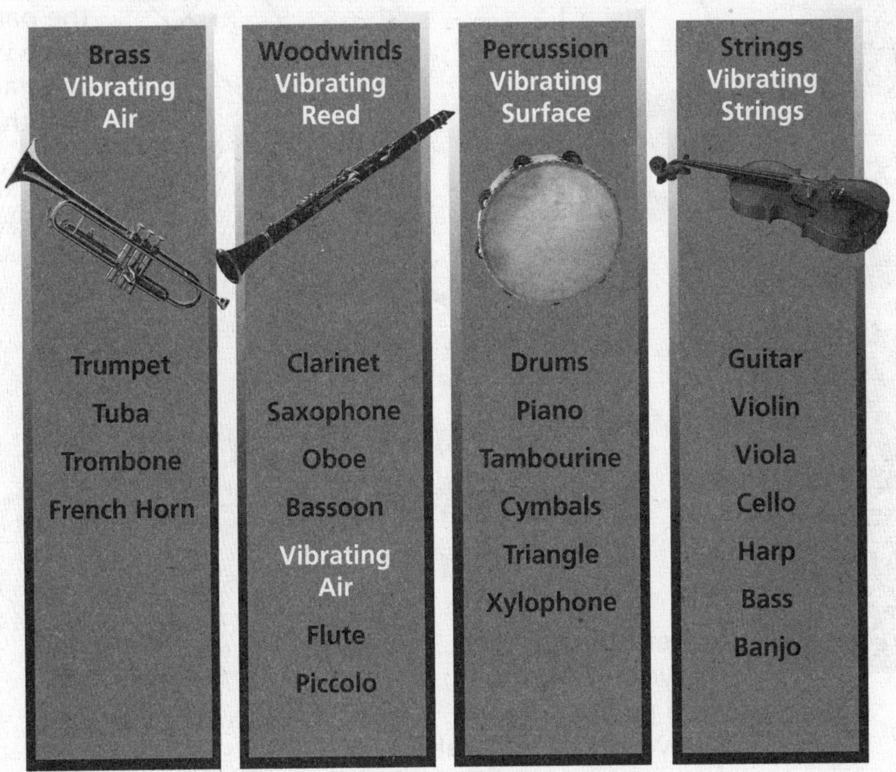

Brass
Vibrating
Air
Trumpet
Tuba
Trombone
French Horn
Woodwinds
Vibrating
Reed
Clarinet
Saxophone
Oboe
Bassoon
Vibrating
Air
Flute
Piccolo
Percussion
Vibrating
Surface
Drums
Piano
Tambourine
Cymbals
Triangle
Xylophone
Strings
Vibrating
Strings
Guitar
Violin
Viola
Cello
Harp
Bass
Banjo

Answer the following questions about the picture above.

1. What part of an oboe vibrates to cause the sound?

____

2. What part of the triangle vibrates to cause the sound?

____

3. What part of the guitar vibrates to cause the sound?

____

4. How are the trumpet and the flute similar?

____

Name_________________________ Date____________

Sound Waves

The picture below shows how the sound waves of a vibrating violin string travel. When the string vibrates, the molecules in the air first crowd together, and then they spread apart as the sound waves travel away from it.

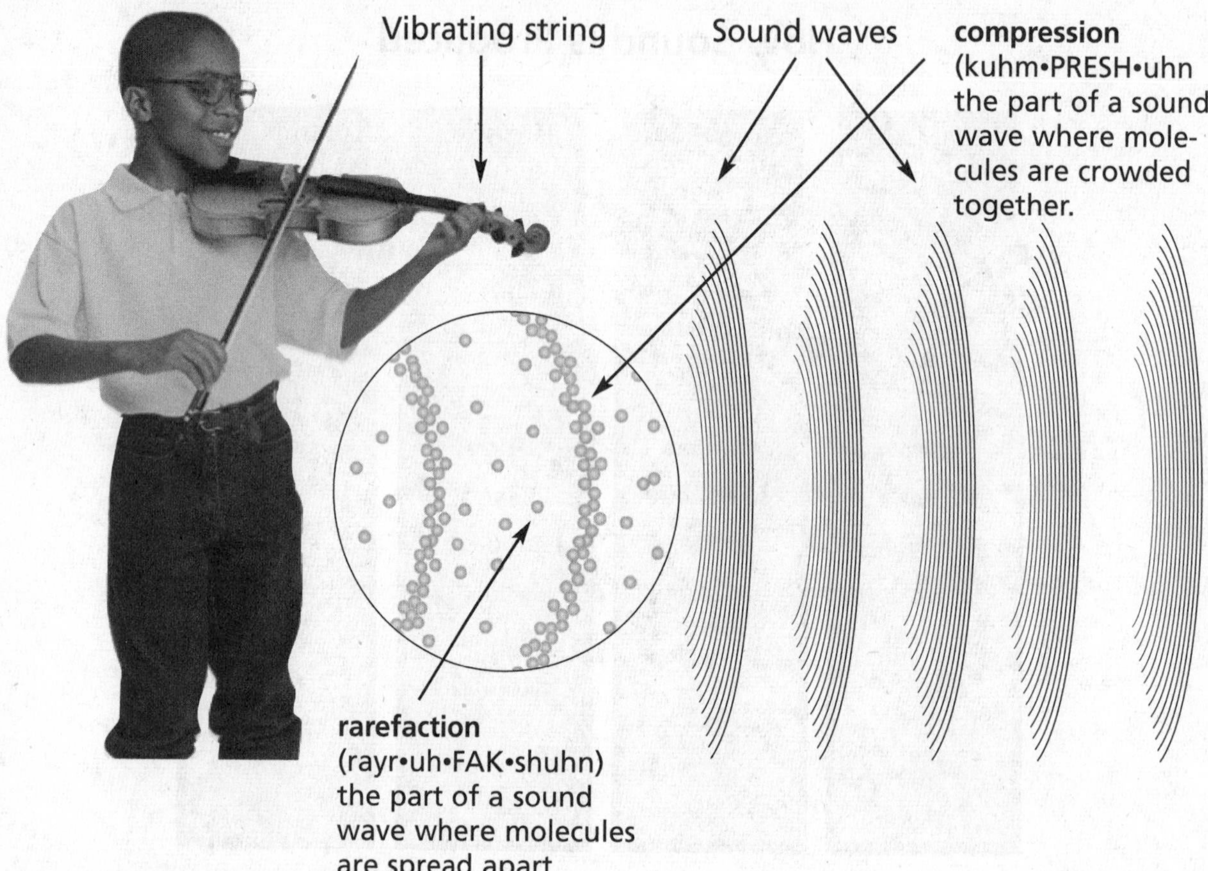
Vibrating string
Sound waves
compression
(kuhm•PRESH•uhn
the part of a sound wave where molecules are crowded together.
rarefaction
(rayr•uh•FAK•shuhn)
the part of a sound wave where molecules are spread apart.

Answer the following questions about the picture above.

1. Which part of the sound wave has molecules that are tightly packed together?

____

____

2. Which part of the sound wave has molecules that are spread apart?

____

3. What did the student do to produce the sound wave?

____

____

Name ______________________ Date __________

Lesson Vocabulary

Lesson 4

Sound Waves

Fill in the blanks.

Vocabulary
surface
ear
matter
vibration
rarefaction
sound wave
compression

1. A rapid back-and-forth movement is called a(n) ______________.

2. Anything that has mass and takes up space is ______________.

3. A percussion instrument makes sound with a vibrating ______________.

4. The part of a sound wave where the molecules are crowded together is called ______________.

5. Sound waves are collected in the human ______________.

6. A vibration that spreads away from a vibrating object is a(n) ______________.

7. The part of a sound wave where the molecules are spread apart is called ______________.

Answer each question.

8. What sounds do you recognize as warnings or alerts?

____

____

9. Explain how sound made by a vibrating object travels.

____

____

____

____

____

____

Name________________________________ Date______________

Cloze Test

Lesson 4

Sound Waves

Vocabulary

sound wave	smallest	matter	molecules
gases	vibrate	ear	directions

Fill in the blanks.

Sound is a vibration that travels through ____________________. Some types of matter are made of pieces too small to be seen, called ____________________. Molecules are the ____________________ pieces that matter can be broken into without changing the kind of matter. When a string vibrates, it makes molecules of gases in the air next to it ____________________. A vibration that spreads away from a vibrating object is a ____________________. It carries the energy from the vibrating object outward in all ____________________. When sound waves reach your ____________________, they make the parts inside vibrate. Since air is a mixture of ____________________, you may conclude that sound can travel through gases.

Name_____________________________________ Date_______________

Pitch and Loudness

Fill in the blanks.  Reading Skill: **Cause and Effect** - questions 2, 3, 7, 11, 13

What Is Pitch?

1. The term ______________________ refers to how high or low a sound is.
2. A shorter string vibrates faster and produces a(n) ______________________ than a longer string.
3. The pitch of a vibrating string is related to its ______________________ as well as its length.
4. The pitch of a person's voice varies with the length and thickness of the ______________________.
5. Scientists use a(n) ______________________ to picture sound waves and compare them.
6. The number of times an object vibrates per second is its ______________________.
7. A lower frequency produces a(n) ______________________ pitch; a higher frequency produces a(n) ______________________ pitch.
8. The unit for measuring frequency is the ______________________, which is abbreviated ______________________.

What Is Volume?

9. The back-and-forth distance that air molecules vibrate is based on how much ______________________ the sound wave carries.
10. A(n) ______________________ sound has more energy than a(n) ______________________ sound.
11. On an oscilloscope, a loud sound produces a(n) ______________________ wave than a soft sound does.
12. The unit for measuring volume is the ______________________.

Name________________________________ Date______________

Fill in the blanks.

How Is Sound Recorded?

13. In the first step of recording, sound waves cause vibrations in the __________________ of a(n) __________________, producing an electric current.

14. In the second step, a(n) __________________ makes the current stronger.

15. Finally, the electric current arranges __________________________ into patterns on a blank tape.

16. On a(n) ____________________, sound is stored by a computer code, not in magnetic patterns.

Name ______________________ Date ____________

Interpret Illustrations

Lesson 5

What Is Volume?

The pictures below show the differences in how a sound wave would look on an oscilloscope if the sound was low pitched, high pitched, loud, or soft. The length between each wavelength corresponds to the pitch of a sound and the height of each wavelength corresponds to the volume of a sound.

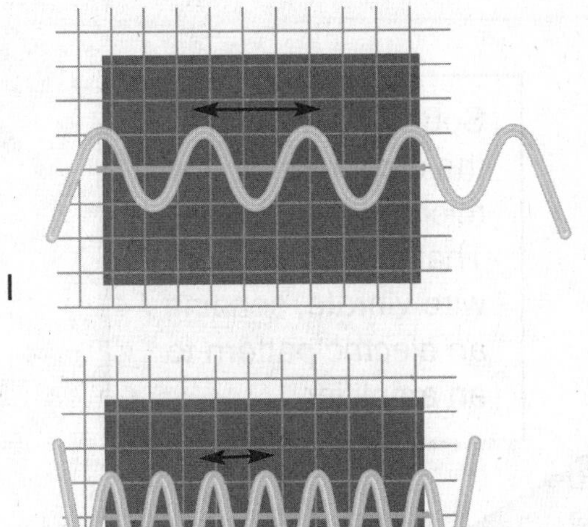

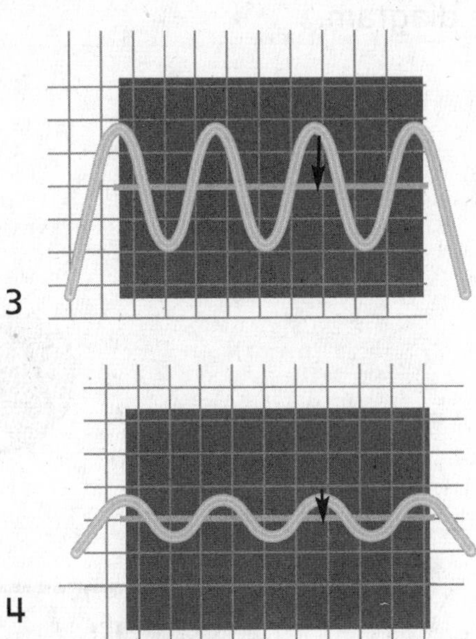

Answer the following questions about the picture above.

1. Which of the oscilloscope pictures shows the highest pitched sound?

____

2. What is the main difference between the sounds shown in pictures 1 and 2, pitch or volume?

____

____

3. The distance between the arrows in pictures 1 and 2 corresponds to what?

____

____

4. The distance between the arrows in pictures 3 and 4 corresponds to what?

____

____

Name ______________________ Date ____________

Interpret Illustrations

Lesson 5

How Is Sound Recorded?

This diagram illustrates the inside of a microphone, showing how sound waves produce an electric current that transmits sound. To understand a diagram like this, read the labels and follow the leader lines to identify each part of the object. Then read the description of what happens in a microphone. Relate each step to the diagram.

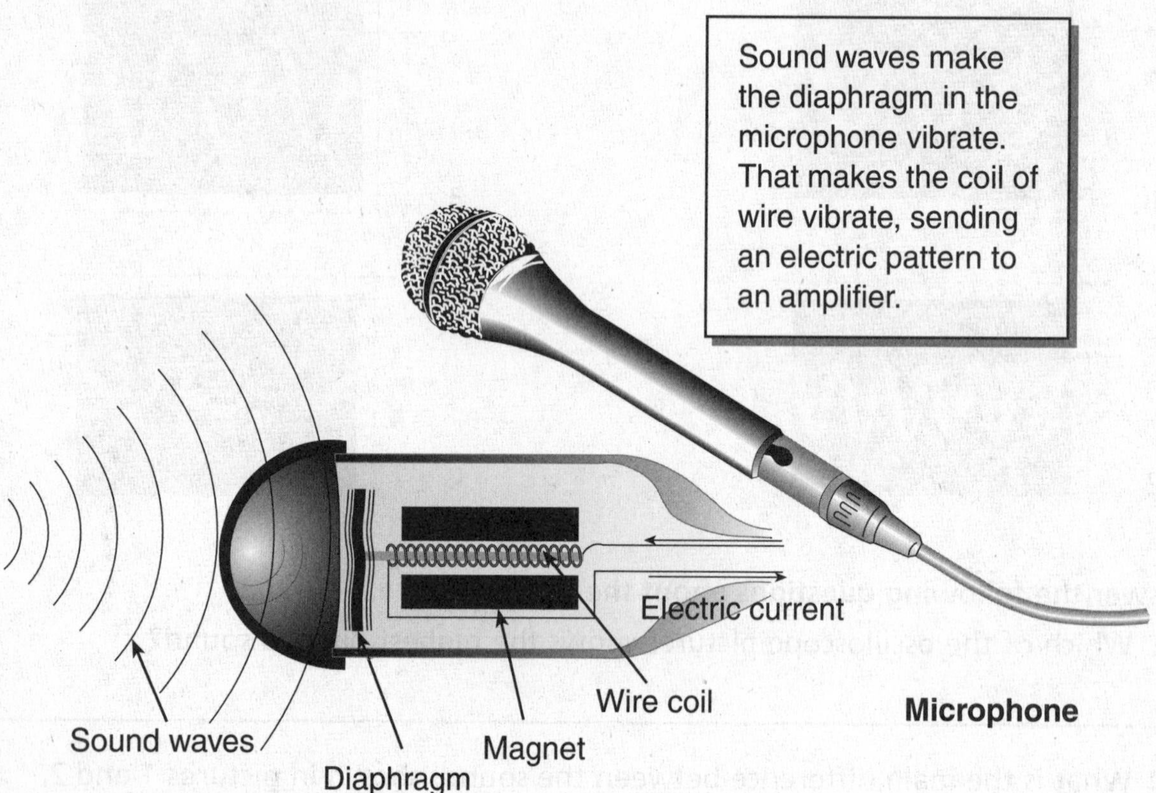

Sound waves make the diaphragm in the microphone vibrate. That makes the coil of wire vibrate, sending an electric pattern to an amplifier.
Electric current
Wire coil
Microphone
Sound waves
Magnet
Diaphragm

Answer these questions about the diagram above.

1. A microphone changes sound waves into a(n) ______________.
2. What do sound waves strike when they reach the head of the microphone? ______________
3. The vibrations of the diaphragm cause vibrations in the ______________.
4. An electric current is produced by the vibrating coil and the nearby ______________.
5. The electric pattern is sent to a(n) ______________.

Name ______________________ Date ____________

Lesson Vocabulary

Lesson 5

Pitch and Loudness

Vocabulary

- frequency
- lower
- decibels
- volume
- pitch
- oscilloscope
- Hertz
- magnetic particles

Fill in the blanks.

1. High and low are words that describe the ________________ of a sound.
2. The more energy a sound wave carries, the greater the ________________.
3. A thicker guitar string produces a(n) ________________ sound than a thinner string because it vibrates more slowly.
4. An instrument that "pictures" sound waves is a(n) ________________.
5. Higher pitched sound waves have a greater ________________.
6. Frequency is measured in units called ________________.
7. A sound's loudness is measured in ________________.
8. Blank recording tapes are coated with scrambled ________________.

Answer each question.

9. How is a drum different from a guitar? Compare the way they make sounds.

____

____

____

10. Would you keep a magnet in the same drawer as you keep your tapes? Why or why not?

____

____

Name ______________________________ Date ______________

Cloze Test

Lesson 5

Pitch and Loudness

Vocabulary

hertz	decibels	pitch	frequency
vibrations	related	oscilloscope	

Fill in the blanks.

You can't see sound waves, but scientists study them with a(n) ________________. This device makes a "picture" of sound waves. ________________ is the number of times an object vibrates per second. Frequency describes ________________ and sound waves. ________________ describes how your brain interprets a sound. Frequency and pitch are ________________: the higher the frequency, the higher the pitch; the lower the frequency the lower the pitch. Frequency is measured in units called ________________. Volume, or how loud or soft a sound is, is measured in units called ________________.

Reflection and Absorption

Fill in the blanks.  Reading Skill: **Cause and Effect** - questions 3, 5, 10, 15, 16, 17, 20

Do Sounds Bounce?

1. When a sound wave hits a surface, some of its energy bounces, or ____________, off the surface.
2. Some energy from the sound wave goes into the surface in the process called ____________.
3. How much of the sound is reflected or absorbed depends on the kind of material of the ____________.
4. A(n) ____________ surface absorbs more sound than a hard surface.
5. In a concert hall, too much ____________ causes a hollow, empty sound.
6. When a music hall was built in the 1870s, people's clothing ____________ more sound than did styles in the 1930s.

What Is an Echo?

7. A reflected sound wave is called a(n) ____________.

How Fast Is Sound?

8. In air at room temperature, sound waves travel at a speed of ____________ per second.
9. Sound waves generally travel faster in ____________ than in ____________ and ____________.
10. The speed of sound waves depends largely on the ____________ of the material.

11. Temperature affects the speed of sound more in ________________ than in liquids and solids.

What Can Echoes Do?

12. Sonar uses ________________ to detect faraway objects.

13. To measure ocean depths, a sonar technician times how long sound waves take to bounce off distant objects and ________________.

14. Animals like whales and bats use a form of sonar called ________________ to find their way or locate food.

How Do Moving Sounds Change?

15. As a siren or train whistle moves toward you, the pitch of the sound gets ________________.

16. When a sound source moves toward you, the ________________ of the sound increases.

17. As the sound source moves away, the sound waves spread, decreasing the ________________ and lowering the ________________.

18. The change in frequency and pitch as a source of sound moves toward or away from you is known as the ________________.

What Is Fundamental Frequency?

19. The lowest frequency of any sound is its ________________.

20. The blend of the fundamental frequency and the ________________ produced gives each sound its own ________________.

21. Bridges have collapsed as a result of ________________.

Name____________________________ Date______________

Interpret Illustrations

Lesson 6

What Can Echoes Do?

This diagram shows two ways sound waves are used to locate objects and measure their distance. Human technicians use sonar technology to measure depths in different parts of the ocean. Whales use a form of sonar called echolocation. To understand the diagram, look carefully at the drawing and read the labels.

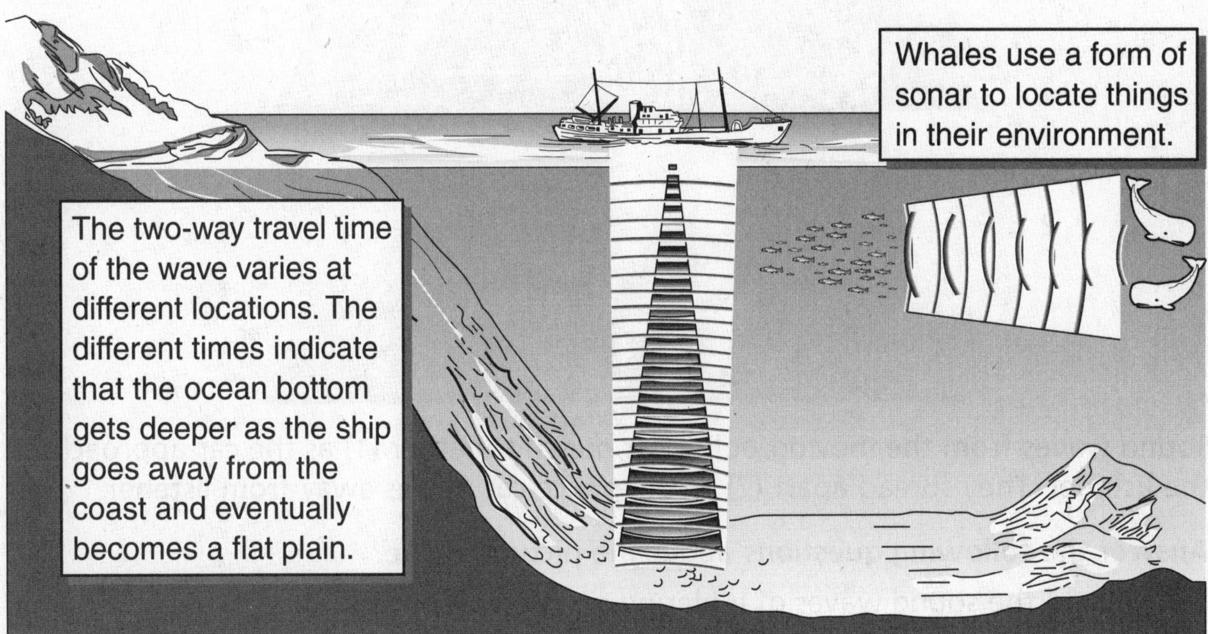

Whales use a form of sonar to locate things in their environment.
The two-way travel time of the wave varies at different locations. The different times indicate that the ocean bottom gets deeper as the ship goes away from the coast and eventually becomes a flat plain.

Answer these questions about the diagram above.

1. What is sonar used for on the ship?

____

2. How can scientists on the ship tell that the depth of the ocean bottom changes as the ship moves away from the coast?

____

3. What happens to the travel time of the wave when the ship is over the flat plain?

____

4. How is the whale using its form of sonar?

____

Name_______________________________ Date_____________

Interpret Illustrations

Lesson 6

How Do Moving Sounds Change?

The pictures below show how the pitch of a sound wave changes as it approaches and moves away from a listener.

2

Sound waves from the moving police car bunch together (1) as the car approaches the listener. They spread apart (2) as the police car moves away from listener.

Answer the following questions about the picture above.

1. Would the sound waves of a blaring siren on a police car bunch together or spread apart as it approaches the listener?

____

2. Would the pitch of the siren get higher or lower as the police car approaches the listener? Why?

____

____

3. Would the sound waves bunch together or spread apart as the police car moves away from the listener?

____

4. Would the pitch of the siren get higher or lower as the police car moves away from the listener? Why?

____

____

Name ____________________ Date __________

Reflection and Absorption

Fill in the blanks.

Vocabulary

resonance
echolocation
fundamental frequency
absorption
reflection
overtone
echo
quality
doppler effect

1. The disappearance of a sound wave into a surface is called ______________.
2. The bouncing of a sound wave off a surface is called ______________.
3. The lowest frequency at which a string vibrates is called ______________________.
4. ______________ is when strings are vibrating at higher frequencies at the same time as the fundamental frequency.
5. The build-up of vibrations at natural frequency is called ______________.
6. Many animals use a form of sonar called ______________.
7. A change in frequency and pitch as sound moves toward and away is called the ______________.
8. A reflected sound wave is called a(n) ______________.
9. ______________ depends on the blend of fundamental frequencies and overtones.

Answer this question.

10. What is the Doppler effect?

____

____

____

____

Name ______________________________ Date ______________

Cloze Test

Lesson 6

Reflection and Absorption

Vocabulary

dolphin	prey	sound waves	sight
echo	echolocation	ears	

Fill in the blanks.

Sonar uses ____________________ to detect objects far away. A form of sonar called ____________________ helps animals locate things around them. Bats use this instead of ____________________ to navigate. They send out squeals toward their ____________________. Bats pick up the reflected sounds using their large ____________________. Another animal that uses sonar to locate objects is the ____________________. You also use reflected sound when you shout hello and make a(n) ____________________.

Name ______________________ Date __________

Sound Energy

Circle the letter of the best answer.

1. The disappearance of a sound wave into a surface is called

a. reflection. **b.** absorption.
c. sonar. **d.** energy.

2. The part of a sound wave where molecules are crowded together is called the

a. rarefaction. **b.** vibration.
c. matter. **d.** compression.

3. All sound waves are created by

a. vibrations. **b.** instruments.
c. plucking a string. **d.** striking a surface.

4. The part of a sound wave where molecules are spread apart is called the

a. compression. **b.** rarefaction.
c. loudness. **d.** tone.

5. Frequency is measured in units called

a. waves. **b.** hertz.
c. radio waves. **d.** decibels.

6. Volume is measured in units called

a. frequency. **b.** loudness.
c. hertz. **d.** decibels.

Name______________________________ Date____________

Chapter Vocabulary

Chapter 15

Circle the letter of the best answer.

7. A vibration that spreads away from a vibrating object is a

a. hertz. **b.** compression.

c. sound wave. **d.** frequency.

8. The bouncing of a sound wave off a surface is called

a. reflection. **b.** absorption.

c. compression. **d.** rarefaction

9. The blend of the fundamental frequencies and overtones gives each sound its own

a. diaphragm. **b.** vibration.

c. quality. **d.** magnet.

10. When a sound wave is absorbed, its sound energy is changed into

a. electrical energy. **b.** light energy.

c. mechanical energy. **d.** heat energy.

11. To hear an echo, you must have

a. a mountain. **b.** a shower stall.

c. a good reflective surface. **d.** a high-tech recording system.

12. When a string vibrates at the fundamental frequency, sections of the string vibrate at a higher frequency called a(n)

a. reflection. **b.** overtone.

c. diaphragm. **d.** hertz.

Name ______________________ Date ____________

Chapter Summary

Chapter Summary

1. What is the name of the chapter you just finished reading?

2. What are four vocabulary words you learned in the chapter? Write a definition for each.

3. What are two main ideas that you learned in this chapter?

Name________________________ Date____________

Chapter Graphic Organizer
Chapter 16

Light Energy

You can use a compare and contrast table to show how two or more items are alike and how they are different. Look at the example shown below for primary colors and primary pigments. The two shaded boxes list the two items being compared. The first column of the table lists the characteristics of the items being compared.

Characteristic	Primary Color	Primary Pigment
Number	3	3
Colors	red, green, blue	magenta, cyan, yellow
Color when mixed together equally	white	black
Refers to	colors of light	colors of pigments, such as paint

Make a compare and contrast table for concave mirrors and convex mirrors.

Characteristic	Concave Mirrors	Convex Mirrors
Curve of shiny surface		
Inside or outside of soup spoon		
More than one type of image?		
Type of image(s)		

Make a compare and contrast table for infrared light and ultraviolet light.

Characteristic	Infrared Light	Ultraviolet Light
Wavelength compared to visible light		
Source		
Effect		

Compare and Contrast

Remember that to compare things, you look for ways the things are alike. To contrast them, you look for ways they are different.

In the exercise below, finish each sentence. The word in parentheses ( ) tells you whether to compare or contrast the two things. The first one has been done for you.

1. (compare) My brother is as tall as I am.
2. (contrast) The boys in our class are ______________________.
3. (contrast) Isn't that car ______________________?
4. (compare) Your new house is ______________________.
5. (contrast) Her hair is ______________________.
6. (compare) Our math teacher is ______________________.
7. (contrast) I think this movie is ______________________.
8. (contrast) My mother's cooking is ______________________.
9. (compare) Yesterday's weather was ______________________.
10. (contrast) Your cousin ______________________.
11. (compare) That painting ______________________.
12. (compare) They said our songs ______________________.
13. (contrast) This candidate for mayor ______________________.
14. (contrast) Those jeans ______________________.
15. (compare) Your computer ______________________.

Name ______________________ Date __________

Chapter Reading Skill
Chapter 16

Something in Common

An **analogy** is a special way to compare things that are different by finding something they have in common. For example, a tadpole is a young frog, and a puppy is a young dog. We can say that *tadpole is to frog as puppy is to dog*. That's an analogy!

Read each example below. Think about the relationship between the first two boldface words. **Then circle the correct word to complete each analogy.**

1. **Ears** are to **sound** as **eyes** are to ______________.

   odors **light** **tastes**

2. **Rug** is to **floor** as **curtain** is to ______________.

   chair **bed** **window**

3. **Elephants** are to **herd** as **geese** are to ______________.

   school **flock** **pride**

4. **Wing** is to **plane** as **sail** is to ______________.

   boat **car** **train**

5. **Evergreen** is to **tree** as **orchid** is to ______________.

   weed **fruit** **flower**

6. **Foot** is to **sneaker** as **hand** is to ______________.

   ring **glove** **blanket**

7. **Knife** is to **knives** as **life** is to ______________.

   living **lived** **lives**

8. **Duet** is to **two** as **trio** is to ______________.

   four **three** **five**

Name_________________________________ Date_______________

Lesson Outline

Lesson 7

Light and Mirrors

Fill in the blanks.  Reading Skill: **Compare and Contrast** - questions 5, 11, 14, 15

Can You See Without Light?

1. The Moon shines only because it reflects ____________________.
2. We cannot see the dark half of the Moon because ________________________________.
3. Light is a means of transferring ____________________ between points.
4. All the objects we can see either give off their own light or ____________________ from another source.
5. All these objects produce their own light in different ways that involve heat:
   a. ____________________ heat the Sun
   b. ____________________ heat a burner flame
   c. ____________________ heats the wire of a light bulb
6. Molecules in a hot material move swiftly and collide, giving off some energy as ____________________.
7. Any light source converts energy of some kind into ____________________.

How Does Light Travel?

8. The shadows that light casts suggest that light usually travels in ____________________.
9. If light passes from one substance into another, it usually ____________________.
10. A beam of light that is not disturbed or bent moves in a straight line as a(n) ____________________.

Name ______________________________ Date ____________

Lesson Outline

Lesson 7

Fill in the blanks.

How Does Light Bounce Off Objects?

11. Light rays that reflect off ____________________ scatter in many directions, while light rays that reflect off a flat, polished surface create a mirror image.

12. According to the ____________________, the angle between an incoming light ray and the surface is ____________________ the angle between the surface and a reflected light ray.

13. Light rays that bounce off a mirror or other shiny surface reflect a picture, or ____________________.

How Do Curved Mirrors Form Images?

14. A mirror that curves in on the shiny side is ____________________, while a mirror that curves out on the shiny side is ____________________.

15. The image in a flat mirror is ____________________ and reversed left to right. The image formed by a ____________________ mirror may be right-side up or upside-down. It may be large or small, depending on how far the object is from the mirror.

How Do Convex Mirrors Work?

16. A convex mirror produces an image that is much ____________________ than the object.

17. Convex mirrors are used as side rearview mirrors in cars because they give a(n) ____________________ view.

Name ______________________________ Date ____________

Interpret Illustrations
Lesson 7

How Does Light Travel?

How does light travel? This drawing shows a light bulb casting the shadow of a microscope onto a wall. If you study the diagram and the paths of the light rays, you can discover for yourself the kind of path that light rays follow. The light rays are lettered so you can identify them.

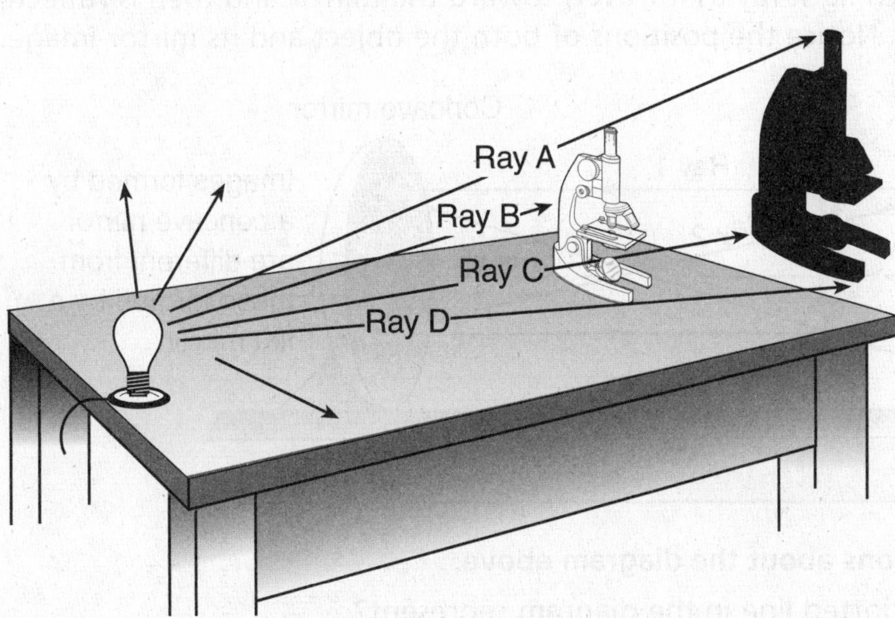

Light casting a shadow suggests that light travels in straight lines.

Answer these questions about the diagram above.

1. Which light ray passes above the top of the microscope and falls onto the wall? ________________
2. A light ray is stopped by the microscope. Which is it? ________________
3. Which light rays pass to the side of the microscope and light up the wall?

   ____
4. Explain in your own words why all the light rays don't go around the microscope and reach the wall.

   ____

   ____
5. What is the result of light rays being stopped by the microscope?

   ____

Name ______________________ Date ____________

Interpret Illustrations

Lesson 7

How Do Curved Mirrors Form Images?

A concave mirror is one that curves inward on the shiny side. This diagram shows how such a curved mirror forms an image. To understand the diagram, notice the arrows that show the paths of two light rays as they reflect off the curved surface. Trace the path of each light ray as it travels toward the mirror and then is reflected back to the observer. Notice the positions of both the object and its mirror image.

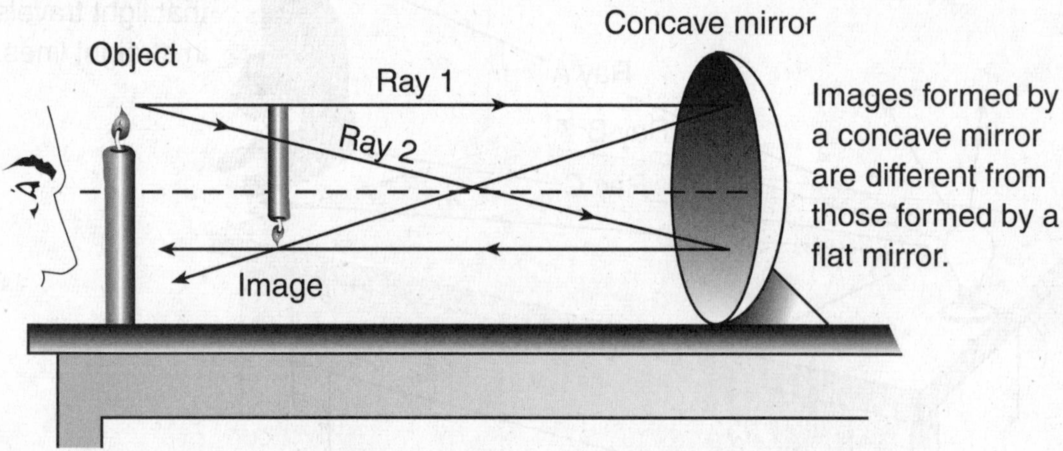

Images formed by a concave mirror are different from those formed by a flat mirror.

Answer these questions about the diagram above.

1. What does the dotted line in the diagram represent?

   ____

2. How does light ray 1 travel toward the mirror?

   ____

   How does it reflect off the curved surface? ______________

3. How does light ray 2 travel toward the mirror?

   ____

   How does it reflect off the curved surface of the mirror?

   ____

4. How does the size of the image compare with the size of the original object?

   ____

5. In what other way is the mirror image different from the original object?

   ____

Name__________________________ Date______________

Lesson Vocabulary
Lesson 7

Light and Mirrors

Fill in the blanks.

Vocabulary
concave
Law of reflection
energy
light ray
chemical
reflects
convex

1. We are able to see most things only because light ________________ off them.
2. Light and sound are both means of transferring ________________ between two points.
3. A straight beam of light is called a ________________.
4. Some chemical reactions produce a(n) ________________ form of light.
5. The ________________ states the angle between an incoming light ray and a surface equals the angle between the reflected light ray and the surface.
6. A(n) ________________ mirror is like the inside of a spoon.
7. A(n) ________________ mirror always forms a reduced, upright image.

Answer each question.

8. You want to check out what is behind you using a mirror. How would knowing about the law of reflection help you position the mirror?

____

____

____

____

____

9. How does light behave like a wave?

____

____

____

____

Name ________________________ Date ____________

Cloze Test
Lesson 7

Light and Mirrors

Vocabulary

concave mirrors	reversed	curve	shiny
convex mirrors	rearview	image	

Fill in the blanks.

Light rays bouncing off a mirror reflect a picture called a(n) ________________. Although reflections in a flat mirror look real, they are ________________ from right to left. Telescopes often use ________________. They ________________ in on the shiny side. Mirrors that curve out on the ________________ side are convex mirrors. You've probably seen ________________ used as a store's security system. Convex mirrors are also used as side ________________ mirrors in cars.

Name________________________ Date____________

Lesson Outline
Lesson 8

Light and Lenses

Fill in the blanks.  Reading Skill: **Compare and Contrast** - questions 1, 7, 9, 12, 13, 17

What Can Light Pass Through?

1. Light easily passes through ________________ materials but is completely blocked out by ________________ materials.
2. Materials that let some light through, but give a blurry view, are called ________________.
3. Light waves usually vibrate in all directions, but ________________ allows only light vibrating in a certain direction to pass through.
4. In sunglasses, polarizing materials block ________________ and other light that vibrates sideways.
5. Self-tinting glasses get darker or lighter because of a chemical containing silver particles that react to ________________.

How Can Light Rays Be Bent?

6. The bending of light rays as they pass from one substance into another is called ________________.
7. When light passes into a denser substance it ________________.
8. A substance made of material that is more tightly packed together is ________________ than one that is less tightly packed together.
9. If light enters a denser material directly, it continues in a straight path, but if it enters at an angle it is ________________ into a new direction.
10. The amount of refraction ________________ as the incoming angle gets shallower.

Name ______________________ Date ____________

Lesson Outline
Lesson 8

Fill in the blanks.

How Do Lenses Work?

11. A curved piece of transparent material that refracts light to make an image is a(n) ____________.

12. The surfaces of a(n) ____________ lens curve outward, while the surfaces of a(n) ____________ lens curve inward.

13. A convex lens refracts light rays ____________, while a concave lens refracts light rays ____________.

How Does the Eye Work?

14. The lens in a human eye is a(n) ____________lens.

15. The lens in your eye casts an image onto a tissue called the ____________, which converts it into nerve signals that go to the ____________.

16. In the eye, light is refracted first by the ____________ and then enters the eye through the pupil and travels to the ____________.

17. In a(n) ____________ eye, the lens focuses images short of the retina, while in a(n) ____________ eye, the image is focused behind the retina.

Name ____________________ Date ____________

Interpret Illustrations
Lesson 8

How Does the Eye Work?

This diagram shows the different parts of the eye, which together allow you to see images. Follow the numbered labels to trace the path of light as it passes through the eye to form an image on the retina. Notice where the light rays are bent, or refracted, when they pass through different structures.

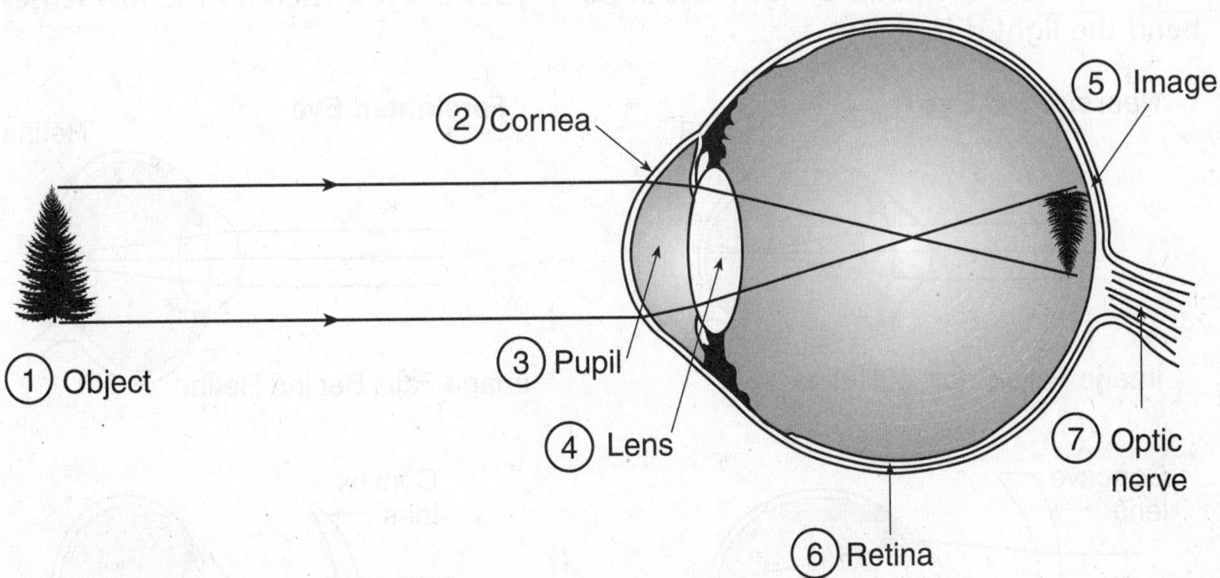

Answer these questions about the diagram above.

1. When light reaches the eye, it is refracted first as it passes through the ____________.
2. The opening that lets light enter the eye is the ____________.
3. Light then passes through the ____________ and is bent still more, forming an image on the retina.
4. The nerve that carries images to the brain is called the ____________.

Name ______________________ Date ____________

Interpret Illustrations

Lesson 8

How Glasses and Contact Lenses Work

These diagrams show how different kinds of lenses can correct certain kinds of vision problems. The lenses bend light rays so that the images fall on the retina. Carefully trace the paths of light rays in both types of eyes. Then notice how lenses bend the light differently.

Nearsighted Eye

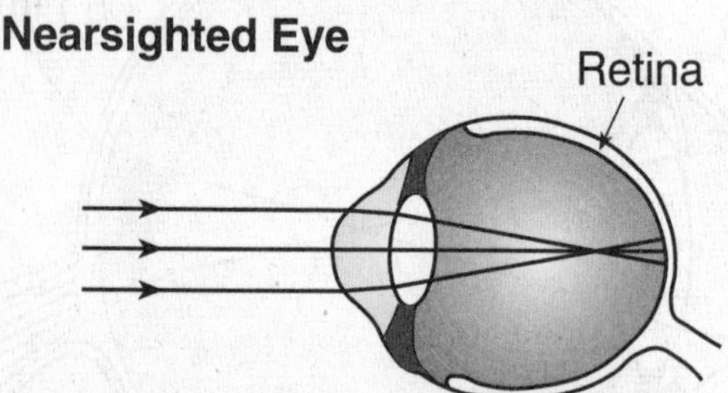

Image Falls Short of Retina

Farsighted Eye

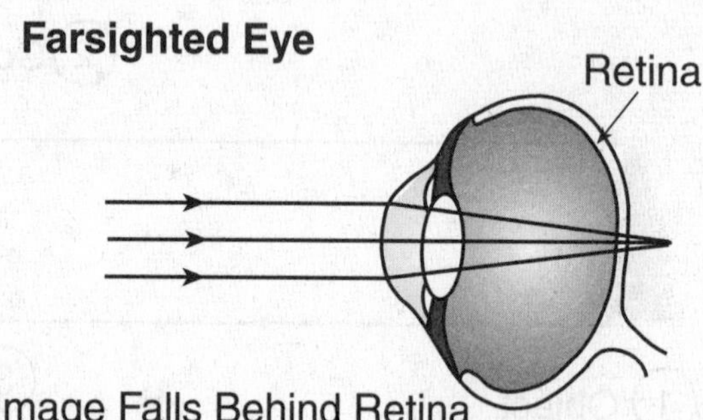

Image Falls Behind Retina

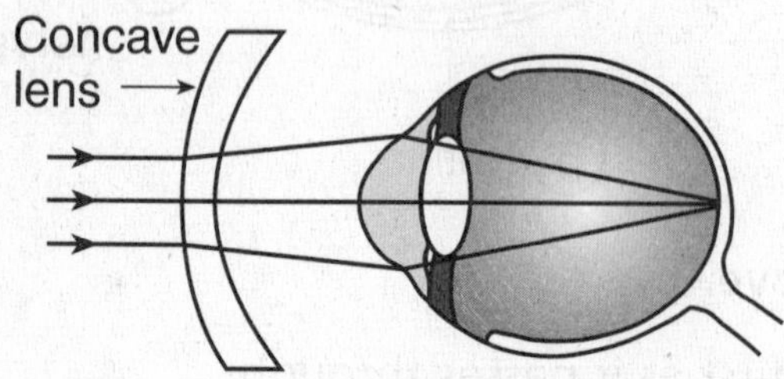

Lens allows image to fall on retina.

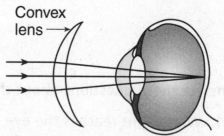

Lens allows image to fall on retina.

Answer these questions about the diagram above.

1. Which type of eye appears shorter from lens to retina?

____

2. In a farsighted eye, does the image fall in front of the retina or behind the retina? ______________________________

3. If a nearsighted person doesn't wear glasses or contact lenses, where does the image fall in the eye? ______________________________

4. Look at the path of the light rays passing through the corrective lens for a nearsighted eye. Does the lens focus the light rays more sharply or spread them out? ______________________________

5. A(n) ______________ lens is used to correct nearsightedness.

Name ______________________ Date ____________

Lesson Vocabulary

Lesson 8

Light and Lenses

Vocabulary

- cornea
- translucent
- opaque
- transparent
- refraction
- convex lens
- polarization
- concave lens

Fill in the blanks.

1. A(n) ________________ object will allow no light rays through.
2. The bending of light rays as they pass from one substance to another is called ________________.
3. When only one direction of light vibrations can pass through a material it is called ________________.
4. A lens that forms images by refracting light rays apart is a(n) ________________.
5. A lens that forms images by refracting light rays together is a(n) ________________.
6. Matter that is ________________ will allow some light to pass through.
7. The ________________ of your eye refracts the image it receives.
8. You can see clearly through any ________________ object.

Answer each question.

9. How are self-tinting sunglasses different from polarized sunglasses? Which would you wear? Why?

____

____

____

____

10. Explain how lenses correct vision problems.

____

____

____

____

Name ______________________ Date ____________

Cloze Test
Lesson 8

Light and Lenses

Vocabulary

direction	color	transparent	particles
polarization	waves	vibrate	materials

Fill in the blanks.

One interesting way for controlling light is called ________________. Light travels in ________________. Normally these waves ________________ in all different directions. Yet, light can be polarized by some ________________. Only one ________________ of light vibration can pass through these materials. Scientists have also used materials to develop sunglasses that change ________________. The lenses of self-tinting glasses contain very small amounts of a ________________, silver-containing chemical. When struck by bright light, this chemical turns into tiny silver ________________. These particles darken the glass in bright light.

Name ______________________ Date ______________

Lesson Outline

Lesson 9

Light and Color

Fill in the blanks.  Reading Skill: **Compare and Contrast** - questions 3, 17

How Do You Get Color From White Light?

1. A(n) ______________ is a triangular piece of polished glass that refracts white sunlight into a band of colors.
2. Newton called the band of colors a(n) ______________.
3. A prism refracts ______________ light the most, and ______________ light the least.
4. If the spectrum made by a prism were passed through a second prism, the colors would recombine to produce ______________.
5. A rainbow forms when raindrops act both as ______________ and as ______________ to reflect the colors to your eyes.
6. If the Sun is ______________, you may see a rainbow in mist from a hose or fountain.
7. White light is a mixture of the entire ______________ of colors.

How Do Colors Look in Colored Light?

8. A color filter ______________ some colors of light and lets others pass through.
9. A red cellophane filter allows ______________ to pass through but blocks ______________.
10. If you look at a red tomato through a green filter, the tomato will appear ______________.
11. The primary colors of light are ______________, ______________, and ______________.

Name ______________________________ Date ______________

Lesson Outline
Lesson 9

Fill in the blanks.

12. Cells in the ______________ of the human eye react to colors of light.
13. If the retina is struck with equal amounts of red, green, and blue light, we see ______________.

What Happens When Color Is Reflected?

14. A(n) ______________ is a colored substance that reflects some colors and absorbs others.
15. The color of an object is due to the colors that it ______________.
16. A green leaf ______________ green light and ______________ red and blue light.
17. A material that absorbs all colors appears ______________, while one that reflects all colors appears ______________.
18. When you mix colored lights, you ______________ colors until you get white.
19. When you mix paints or pigments, you ______________ colors until you get black.
20. The primary pigments are ______________, ______________, and ______________.

Name________________________ Date____________

How Do Colors Look in Colored Light?

The picture below shows how a red tomato would look if it were looked at through different colored filters. Red, green, and blue are considered to be the primary colors of light.

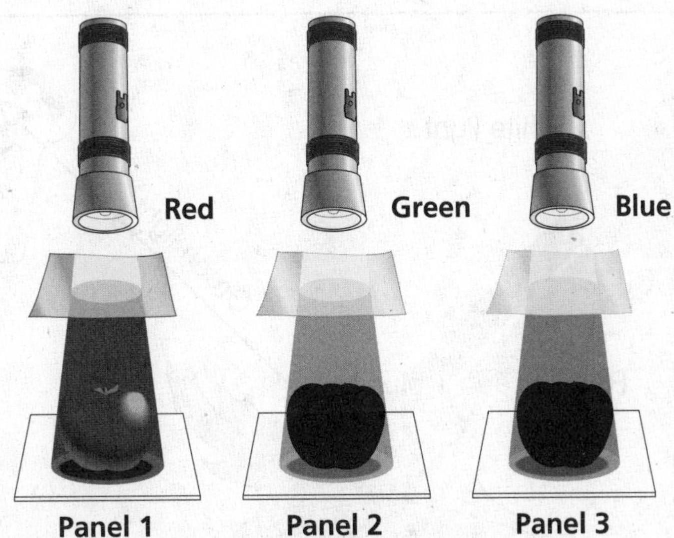

Answer the following questions about the picture above.

1. Why does the red tomato appear to be red under the red filter?

____

____

2. Why does the red tomato appear to be black under the green or blue filters?

____

____

____

3. What colors would a green apple appear to be under each of the filters?

____

____

Name_____________________________ Date______________

Interpret Illustrations

Lesson 9

What Happens When Color Is Reflected?

This diagram shows why different objects are different colors. Pigments in the object reflect or absorb certain colors of light. To understand the diagram, follow the path of the ray of light as it strikes a green leaf and then is reflected back to a person's eye. Notice what happens to the different colors in the light ray.

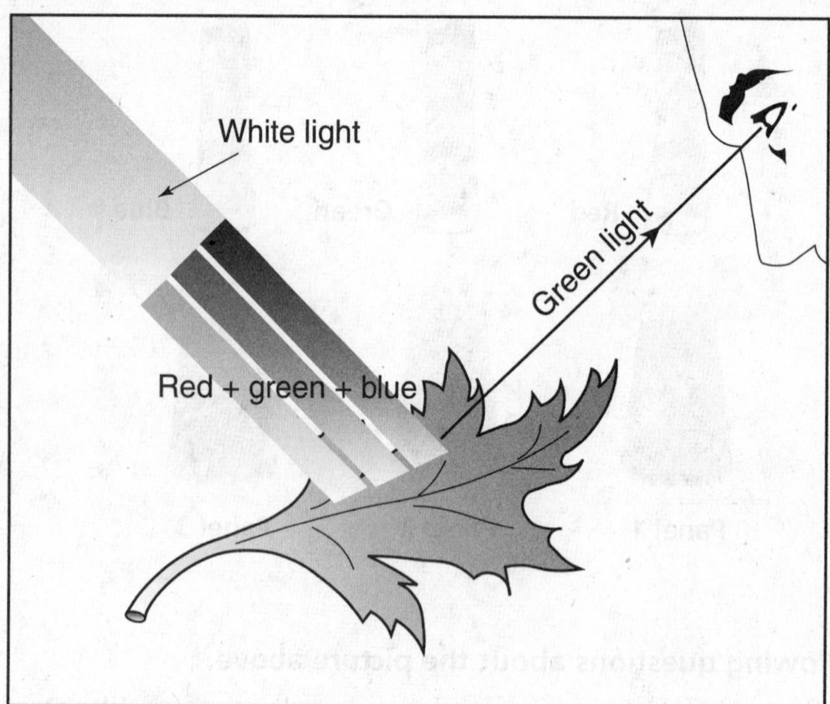

Answer these questions about the diagram above.

1. What color is the ray of light that strikes the leaf? ____________________
2. A ray of white light is actually a blend of red, ______________, and ______________ light.
3. What color or colors of light does the green leaf absorb?

   ____

4. What color or colors of light does the green leaf reflect? ______________
5. Imagine that the light is striking a red rose instead of a green leaf. What color or colors would the rose absorb? ____________________

Name ______________________________ Date ______________

Lesson Outline
Lesson 9

Light and Color

Vocabulary

spectrum
refracted
retina
black
prism
primary colors
white
primary pigments

Fill in the blanks.

1. If you shine a light through a(n) ______________, you can see a rainbow.
2. The ______________ of light are red, green, and blue.
3. A prism will break light apart by color so you can see the whole ______________.
4. An object that reflects all colors will appear ______________.
5. The color violet is ______________ more than any other color in the spectrum.
6. The ______________ are magenta, cyan, and yellow.
7. An object that absorbs all colors will appear ______________.
8. Cells in the eye's ______________ allow us to see colors.

Answer each question.

9. How is mixing the colors of light different than mixing colors of paint?

____

____

____

____

10. What color filter or filters would make a red tomato appear red? Black?

____

____

____

____

Name ______________________ Date ______________

Cloze Test

Lesson 9

Light and Color

Vocabulary

refracts	magenta	rainbow	colors	sunlight
prism	primary	spectrum	drops	

Fill in the blanks.

Sir Isaac Newton discovered that when light goes through a(n) ______________ it produces a band of rainbow colors. He named this color band a(n) ______________. White sunlight is a mixture of many ______________. The prism ______________ each color at a different angle. Water ______________ can also act like prisms. They can break up rays of ______________ into different colors. When this occurs, a(n) ______________ forms in the sky. We can paint these colors ourselves by mixing ______________ pigments. These include yellow, ______________, and cyan.

Name ______________________ Date ____________

Lesson Outline
Lesson 10

Invisible Light

Fill in the blanks.  Reading Skill: **Compare and Contrast** - questions 7, 16

How Do Waves Move?

1. All waves carry ______________ from place to place.
2. Without particles vibrating back and forth in the ______________ that the sound is traveling, the energy of the sound vibration could not travel.
3. Sound waves cannot travel in a(n) ______________, a space where there is no matter.

How Do Light Waves Travel?

4. Electricity and magnetism produce forces of ______________.
5. Light is a form of ______________.
6. The ______________ of a wave is the distance from the crest of one wave to the crest of the next.
7. Electromagnetic waves vibrate back and ______________ across the direction in which light travels.
8. The ______________ of light are different wavelengths.
9. Different wavelengths of light travel through space at ______________.
10. Another theory about how light travels suggests that it moves as bundles of energy called ______________.
11. The human eye cannot see wavelengths of light that are ______________ than red or ______________ than violet.
12. Visible and invisible wavelengths of light together make up the ______________.

Name_______________________________ Date_______________

Fill in the blanks.

Which Wavelengths Are Longer than Red Light?

13. The longest waves in the electromagnetic spectrum are _______________.

14. A radar device uses _______________ to locate objects.

Which Wavelengths Are Shorter than Violet Light?

15. Ultraviolet light can produce _______________ in your body.

16. UV light from the Sun can cause a _______________ as well as some forms of the disease _______________.

17. Earth is protected from much of the Sun's UV light by the _______________.

18. The shortest wavelengths in the spectrum are _______________ and _______________.

What Are Lasers?

19. Unlike light from most sources, a(n) _______________ is a device that produces thin streams of light.

Name____________________________ Date__________

Interpret Illustrations

Lesson 10

How Do Waves Move?

These drawings show how a moving wave affects a floating object. Compare the three drawings and notice how the crest of the wave moves forward.

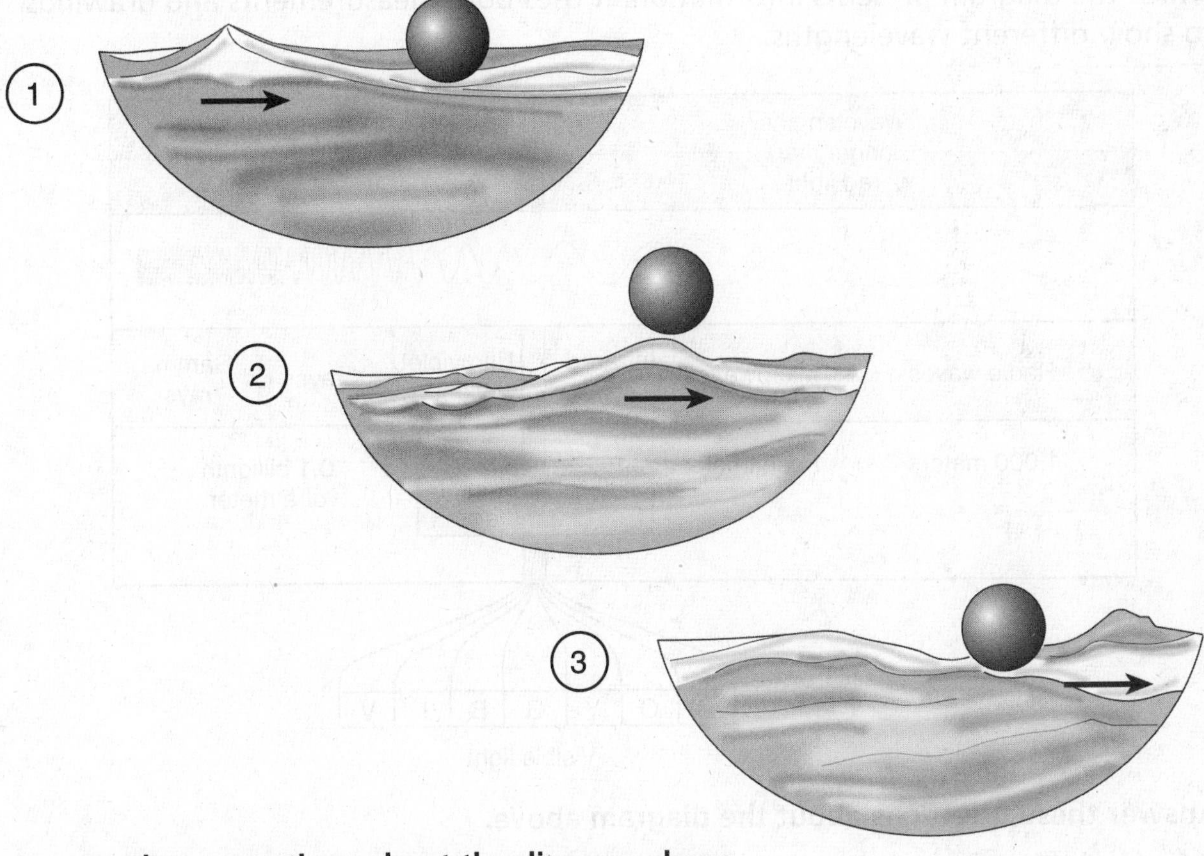

Answer these questions about the diagram above.

1. In which direction is the water wave moving?

____

2. What happens to the ball when the wave crest reaches it?

____

3. What happens to the ball after the wave crest passes it?

____

4. Does the wave move the ball forward or backward? Explain your answer.

____

____

Name ______________________ Date ____________

Interpret Illustrations

Lesson 10

How Do Light Waves Travel?

This diagram shows the kinds of light—visible and invisible—that make up the electromagnetic spectrum. Read this diagram as you would a chart. Notice major categories and the smaller divisions within them. Look at the different ways in which the diagram presents information. It uses both measurements and drawings to show different wavelengths.

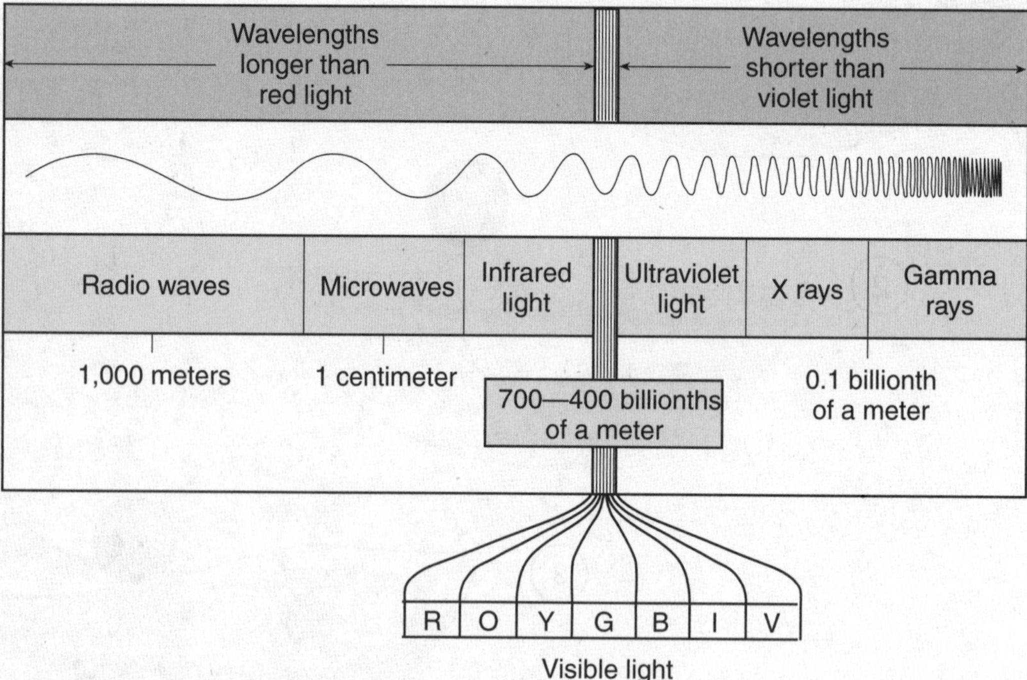

Visible light

Answer these questions about the diagram above.

1. Are infrared waves longer or shorter than red light?

____

2. Are ultraviolet rays longer or shorter than violet light?

____

3. The longest waves in the spectrum are ______________________.

4. Which rays have a longer wavelength—X rays or ultraviolet rays?

____

5. The wavelength of microwaves is ______________________.

6. In visible light, which color has the longest wavelength?

____

Name ______________________ Date __________

Lesson Outline
Lesson 10

Invisible Light

Vocabulary

- radio
- lasers
- matter
- electromagnetic spectrum
- vacuum
- wavelengths
- photons
- electromagnetism

Fill in the blanks.

1. Sound waves cannot travel in a(n) ______________.
2. Electromagnetic waves can travel with or without ______________.
3. We cannot see ______________ of light longer than the length of red light waves.
4. The longest waves of the electromagnetic spectrum are ______________ waves.
5. ______________ refers to forces that come from electricity and magnetism.
6. Scientists believe that light may travel as tiny bundles of energy called ______________.
7. ______________ produce light that does not spread out or become weaker.
8. All of the wavelengths of light, the ones we see and the ones we cannot see, are called the ______________.

Answer each question.

9. What instrument allows scientists to detect forms of light that they cannot see?

____

____

10. List, in order of longer to shorter wavelengths, the different waves of the electromagnetic specturm.

____

____

Name______________________________ Date____________

Cloze Test
Lesson 10

Invisible Light

Vocabulary

light	ranging	echo	signals
photons	electromagnetic spectrum	models	vibrating

Fill in the blanks.

Since James Clark Maxwell's work with electromagnetic energy, scientists have formed another idea of how light travels. Rather than as a smooth __________________ wave, perhaps light travels as tiny bundles of energy. Scientists call the tiny bundles of energy __________________. Waves or photons? Scientists use both __________________ to explain light. Some invisible wavelengths of __________________ are longer than red. Radar, short for radio detecting and __________________, uses radio waves that reflect off objects. This is similar to how animals locate things using the __________________ of sound waves. The longest waves of the ____________________________________ are radio waves. These waves carry __________________ in AM or FM codes.

Name______________________ Date______________

Chapter Vocabulary

Chapter 16

Light Energy

Circle the letter of the best answer.

1. A mirror that produces an upright, reduced image is a
   a. concave mirror.
   b. flat mirror.
   c. highly polished mirror.
   d. convex mirror.

2. A point on a light wave as it ripples outward from its source is a
   a. spectrum.
   b. light ray.
   c. spark.
   d. color pigment.

3. A lens that forms images by refracting light rays together is a
   a. convex lens.
   b. magnetic lens.
   c. concave lens.
   d. laser lens.

4. Concave mirrors
   a. form an image that is larger than the original object.
   b. form many different kinds of images.
   c. form an image that appears to be upside down.
   d. form an image that is smaller than the original object.

5. Materials with this property completely block light from passing through and are called
   a. concave.
   b. convex.
   c. transparent.
   d. opaque.

6. Clear glass is an example of a(n)
   a. transparent material.
   b. opaque material.
   c. mirror.
   d. translucent material.

Name______________________________ Date______________

Chapter Vocabulary

Chapter 16

7. When a material allows only part of the light to pass through it is called

a. opaque. b. pigmented.

c. translucent. d. transparent.

8. If an object appears to be red, it

a. is red.

b. reflects red light and absorbs all the other colors.

c. absorbs red light and reflects all the other colors.

d. has molecules of red which are more active than the other molecules.

9. A triangular piece of cut and polished glass is a

a. mirror. b. lens.

c. prism. d. laser.

10. Which is NOT a primary color of light?

a. brown b. green

c. blue d. red

11. What are the three primary pigments?

a. yellow, red, and blue b. green, blue, and red

c. magenta, cyan, and yellow d. magenta, cyan, and fuchsia

12. All the wavelengths of light, seen or unseen, are called the

a. photons. b. prism.

c. infrared waves. d. electromagnetic spectrum.

13. Laser beams are narrow and

a. blue. b. red.

c. direct. d. weak.

Name ____________________ Date ____________

Unit Vocabulary

Unit F

Meanings and Words

Vocabulary			
echo	inertia	frequency	vibration
transparent	concave	convex	opaque
refraction	weight	action	spectrum

Read each meaning. Then find a word in the Vocabulary Box that fits that meaning, and write it on the line.

Meanings	Words
1. a back-and-forth motion	____________
2. the number of times an object vibrates per second	____________
3. a reflected sound wave	____________
4. tendency of an object to resist a change in its state of motion	____________
5. completely blocking light passage	____________
6. the bending of light rays	____________
7. curves in	____________
8. curves out	____________
9. the force of gravity on any object	____________
10. a band of colors produced when light goes through a prism	____________
11. when one object applies a force to a second	____________
12. letting all light through	____________

Crossword

Read each clue. Write the answer.

Across

1. Unit that measures loudness
4. Unit for measuring frequency
6. Opposite of low
9. Back-and-forth motion
10. Tiny bundle of light energy
11. Bending light rays
13. How high or low a sound is
14. Difference between two sounds of the same pitch
16. Curves in
17. Opposite of high
18. Curves out

Down

2. Bounced-back sound
3. Straight-line beam of light
5. Part of a sound wave where molecules are spread apart
7. Opposite of out
8. A noise
12. Blocks light going through
15. "Picture" of a light source

Name______________________________ Date______________

Unit Vocabulary

Unit F

Find-a-Word

Look across, down, and diagonally to find these hidden words:

ABSORPTION, BLUE, COLOR, CONCAVE, CONVEX, CURVED, CYAN, DECIBEL, EAR, ECHO, ENERGY, EYE, FREQUENCY, GREEN, HEAR, HERTZ, HIGH, IMAGE, LASER, LENS, LOUD, LOW, OPAQUE, PHOTON, PIGMENT, PITCH, PRISM, RAREFACTION, RAY, RED, REFLECTION, SOFT, SOUND, SPECTRUM, TRANSLUCENT, ULTRASONIC, VACUUM, VIBRATION.

R E F L E C T I O N E A R U
R A Y A C O G Q P R I S M L
C L R S H L R V A C U U M T
U O Z E O O E Z Q B L U E R
R W X R F R E Q U E N C Y A
V H I G H A N L E N S Z E S
E I M A G E C O N C A V E O
D A B S O R P T I O N Z S N
P E Z R E D Z P I T C H P I
I S C Y A N Q H Q O O E E C
G O H I Q T S O F T N N C H
M U E Z B Z I T X Z V E T E
E N A X F E L O U D E R R R
N D R J Q X L N N X X G U T
T R A N S L U C E N T Y M Z